TABLE OF CONTENTS

Dedicate To God

Unveiling the Cosmic Question Are We Truly Alone?

BY

WILSON ERUEMULOR

Are We Truly Alone?

The question of whether we are truly alone in the universe is one that has captivated the minds and imaginations of humanity for centuries. While the answer remains elusive, the pursuit of this question has led to remarkable discoveries, profound philosophical contemplation, and a sense of wonder about the cosmos that transcends cultural and geographical boundaries.

Exploring this question has sparked scientific inquiry into the search for extraterrestrial life, leading to the discovery of exoplanets in distant solar systems and the tantalizing possibility of habitable worlds beyond our own. The emergence of astrobiology as a scientific discipline has brought together researchers from diverse fields to investigate the conditions necessary for life to arise and thrive elsewhere in the universe.

Moreover, the question of our place in the cosmos has inspired countless works of art, literature, and film that explore the themes of alien contact, interstellar exploration, and the implications of discovering intelligent life beyond Earth. These imaginative and thought-provoking creations have invited audiences to ponder the potential consequences of making contact with extraterrestrial civilizations and the profound impact such an encounter could have on our understanding of our place in the universe.

In the realm of philosophy, the question of whether we are alone in the universe has prompted contemplation about the nature of consciousness, the ethical considerations of interstellar communication, and the implications of encountering beings whose existence challenges our understanding of life, intelligence, and the cosmos.

Ultimately, the question of whether we are truly alone in the universe serves as a reminder of the boundless mysteries that await us beyond the confines of our planet. It invites us to embrace a spirit of curiosity, exploration, and humility as we continue to seek answers that may forever expand our understanding of the cosmos and our place within it. The quest to unravel this cosmic enigma is a testament to humanity's enduring fascination with the unknown and our unyielding pursuit of knowledge that transcends the limits of Earth.

The age-old human fascination with extraterrestrial life

The age-old human fascination with extraterrestrial life spans cultures, epochs, and disciplines, reflecting a profound and enduring curiosity about the possibility of life beyond Earth. This fascination is deeply rooted in the human psyche and has manifested in a variety of ways throughout history, captivating imagination and inspiring exploration.

Mythology and Folklore: Across cultures, myths and folklore abound with tales of otherworldly beings, celestial visitors, and divine messengers from the stars. These narratives often reflect humanity's awe and wonder at the mysteries of the night sky and the possibility of beings inhabiting distant realms.

Literature and Science Fiction: The exploration of extraterrestrial life has been a prominent theme in literature and science fiction, offering imaginative depictions of alien worlds, civilizations, and encounters. From the pioneering works of Jules Verne and H.G. Wells to modern masterpieces by authors like Arthur C. Clarke and Ursula K. Le Guin, the genre has captivated audiences with thought-provoking narratives that explore the implications of alien contact and the diversity of life in the cosmos.

Scientific Inquiry: The quest to discover extraterrestrial life has driven scientific exploration, from the pioneering work of astronomers and astrobiologists to the search for habitable exoplanets and the investigation of extremophiles on Earth. The development of space telescopes, planetary rovers, and the ongoing search for biosignatures in the cosmos reflect humanity's steadfast pursuit of unraveling the mysteries of life beyond our home planet.

Popular Culture: Extraterrestrial life has permeated popular culture, from blockbuster films and television series to video games and comic books. These portrayals often reflect societal hopes, fears, and dreams regarding the potential for alien contact and the implications of discovering intelligent life beyond Earth.

Philosophical Reflection: The question of extraterrestrial life has sparked profound philosophical contemplation, touching on themes of cosmic purpose, the nature of consciousness, and the ethical considerations of interstellar communication. Philosophers and thinkers have grappled with the implications of the existence of other intelligent beings and the profound impact such a discovery could have on humanity's understanding of its place in the universe.

In sum, the age-old human fascination with extraterrestrial life speaks to our innate curiosity, imagination, and wonder at the vastness of the cosmos. It reflects a universal quest for understanding, connection, and meaning in a universe teeming with possibilities, inviting us to embrace a spirit of exploration and inquiry that transcends the boundaries of Earth.

The importance of the question for humanity

The question of whether we are alone in the universe holds profound importance for humanity on multiple levels, shaping our worldview, scientific pursuits, and philosophical reflections in significant ways:

Cosmic Perspective: Exploring the potential existence of extraterrestrial life encourages humanity to adopt a broader cosmic perspective. It invites us to consider our place in the universe, fostering a sense of humility, wonder, and interconnectedness with the broader tapestry of life that may exist beyond our planet.

Scientific Inquiry: The quest to understand extraterrestrial life drives scientific inquiry, innovation, and exploration. It motivates astronomers, astrobiologists, and planetary scientists to investigate exoplanets, search for biosignatures, and explore the potential for habitable environments beyond Earth. This pursuit pushes the boundaries of human knowledge, technology, and understanding of the cosmos.

Technological Advances: The search for extraterrestrial life has historically spurred technological advancements, from the development of telescopes and space probes to the exploration of extreme environments on Earth. These advancements often have practical applications that benefit society, including in medicine, environmental science, and space exploration technology.

Cultural and Philosophical Reflection: The question of extraterrestrial life provokes deep philosophical reflections on the nature of consciousness, the ethics of interstellar communication, and the implications of encountering other intelligent beings. It inspires creativity, contemplation, and exploration of humanity's place in the universe, igniting conversations that transcend cultural, religious, and national boundaries.

Societal Impact: The potential discovery of extraterrestrial life could have profound societal impact, shaping our collective sense of identity, purpose, and responsibility as inhabitants of a vast and diverse universe. It may foster a greater appreciation for the fragility of life on Earth and underscore the importance of stewardship and cooperation on a global scale.

The question of whether we are alone in the universe holds immense importance for humanity, inspiring us to embrace a holistic perspective of the cosmos, to advance scientific knowledge and technological innovation, and to engage in philosophical and cultural conversations that broaden our understanding of our place in the universe. It invites us to ponder the profound implications of potential discoveries and to consider the far-reaching impact on our collective consciousness and aspirations as inhabitants of a wondrous and enigmatic cosmos.

Setting the stage for a cosmic exploration

Setting the stage for a cosmic exploration involves a multifaceted approach that encompasses scientific, technological, cultural, and philosophical dimensions. Here's how we can set the stage for a grand cosmic exploration:

Scientific Inquiry and Research: Establishing a robust foundation for cosmic exploration begins with scientific inquiry and research. This involves investing in cutting-edge astronomical observations, space-based telescopes, and missions designed to study exoplanets, planetary environments, and astronomical phenomena that may hold clues to the existence of extraterrestrial life.

Astrobiological Investigations: In tandem with astronomical research, fostering interdisciplinary collaborations in the field of astrobiology is crucial. Scientists from diverse disciplines, including biology, chemistry, geology, and astronomy, can work together to investigate the conditions necessary for life to arise and thrive beyond Earth.

Space Exploration Missions: Initiating and supporting ambitious space exploration missions is essential for venturing beyond our home planet. This includes robotic missions to explore the surfaces of other worlds, manned missions to the Moon, Mars, and beyond, and the development of advanced propulsion technologies for interstellar travel.

Technological Advancements: Advancing space exploration capabilities through technological innovation is paramount. This involves developing advanced propulsion systems, life support technologies, habitat designs, and autonomous exploration platforms that can withstand the rigors of deep space and alien environments.

International Collaboration: Collaboration among nations, space agencies, and private enterprises is vital for fostering a global effort in cosmic exploration.

Establishing partnerships for space missions, information sharing, and resource allocation can leverage the unique strengths and capabilities of different countries and organizations.

Public Engagement and Education: Educating and engaging the public in the excitement of cosmic exploration is crucial for garnering support and inspiring future generations of scientists, engineers, and explorers. Outreach programs, educational initiatives, and public events can nurture a sense of wonder and enthusiasm for the cosmos.

Ethical and Philosophical Considerations: Contemplating the ethical and philosophical implications of cosmic exploration is essential. Deliberations on interstellar communication, planetary protection, and the cultural and societal impact of potential discoveries can inform thoughtful and responsible approaches to cosmic exploration.

By embracing this multifaceted approach, we can set the stage for a grand cosmic exploration that ignites the human spirit of curiosity, discovery, and unity, paving the way for a future where humanity ventures boldly into the cosmos, driven by the timeless quest to unravel the mysteries of the universe and seek out new frontiers of exploration.

Ethical considerations in searching for extraterrestrial life

The search for extraterrestrial life raises a host of ethical considerations that merit thoughtful contemplation and responsible action. Here are some key ethical considerations in searching for extraterrestrial life:

Planetary Protection: Maintaining the integrity of celestial bodies and preventing contamination during space exploration missions is essential for preserving the potential for finding extraterrestrial life. Striving to avoid introducing Earth-based organisms to other worlds and safeguarding our own planet from potential contaminants brought back from space are critical ethical imperatives.

Interstellar Communication: Contemplating the ethical implications of interstellar communication with potential extraterrestrial civilizations is a complex and nuanced consideration. Deliberations on the potential impact of communication, the necessity for consensus, and the application of ethical frameworks in interstellar messaging are crucial aspects of this ethical consideration.

Cultural Sensitivity: Recognizing and respecting the cultural, societal, and ethical norms of any potential extraterrestrial civilizations, should they be encountered, is an important ethical consideration. Sensitivity to diverse cultural perspectives and values in the event of interstellar contact is essential for fostering understanding, respect, and cooperation.

Impact on Earth: Considering the societal, cultural, and psychological impact of potential discoveries of extraterrestrial life on Earth is a significant ethical consideration. Evaluating the potential implications for human beliefs, worldviews, and societal structures can inform responsible approaches to managing the societal impact of such discoveries.

Stewardship of the Cosmos: Embracing a sense of stewardship for the cosmos and the potential habitats of extraterrestrial life is an ethical imperative. Contemplating the responsible use of resources, the preservation of celestial environments, and the ethical dimensions of exploration and colonization beyond Earth is important for fostering a sustainable and ethical presence in the universe.

Global Collaboration and Transparency: Fostering open, transparent, and globally collaborative initiatives in the search for extraterrestrial life can promote ethical conduct, information sharing, and the responsible allocation of resources. International engagement and cooperation contribute to ethical decision-making and the advancement of cosmic exploration.

Through thoughtful consideration of these ethical dimensions, stakeholders in the search for extraterrestrial life can work toward responsible and ethical approaches that uphold the integrity of the cosmos, respect the diversity of potential life beyond Earth, and promote harmonious and informed interactions in the event of significant discoveries or encounters.

Astrophysics and the Search for Alien Life

Astrophysics plays a pivotal role in the search for alien life, offering a framework for understanding the conditions, environments, and phenomena that may harbor and support life beyond Earth. Here are several key ways in which astrophysics contributes to the search for alien life:

Identifying Habitable Zones: Astrophysicists utilize their understanding of stellar evolution, planetary dynamics, and habitable zone calculations to identify regions around stars where conditions may be suitable for the existence of liquid water, a key ingredient for life as we know it. This informs the search for exoplanets within habitable zones and guides the selection of targets for further study.

Characterizing Exoplanets: Leveraging techniques such as transit photometry, radial velocity measurements, and direct imaging, astrophysicists work to characterize the atmospheres, compositions, and properties of exoplanets. This helps in determining the potential habitability and the presence of biosignatures such as water vapor, oxygen, and methane that may indicate the presence of life.

Understanding Planetary Habitability: Astrophysics contributes to our understanding of the factors that contribute to planetary habitability, including the role of a planet's size, composition, magnetic field, and geological activity. This allows scientists to assess the potential habitability of exoplanets and planetary bodies within our own solar system.

Investigating Extreme Environments: Astrophysical research into extreme environments, such as extremophiles on Earth, planetary analog sites, and the potential for life in extreme conditions, informs the search for life in environments beyond Earth. This includes exploring the limits of terrestrial life and extrapolating to consider the potential existence of extremophiles on other worlds.

Searching for Extraterrestrial Signals: The field of astrophysics is involved in the search for potential signals of extraterrestrial intelligence, such as radio and optical transmissions. This involves listening for anomalous signals and distinguishing natural astrophysical phenomena from potential artificial sources, contributing to the exploration of the potential for intelligent life beyond Earth.

Informing Space Mission Planning: Astrophysical insights guide the planning and execution of space missions aimed at studying exoplanets, planetary moons, and other environments in the search for extraterrestrial life. This includes the design and operation of telescopes, spectrographs, and other scientific instruments used in the direct study of potential habitable worlds.

Through these contributions and more, astrophysics provides a foundational framework for the search for alien life, informing the scientific inquiry, technological development, and philosophical contemplation that underpin humanity's quest to unravel the cosmic enigma of life beyond Earth.

The role of astrophysics in understanding the universe

Astrophysics plays a crucial role in deepening our understanding of the universe by encompassing a wide array of phenomena, theories, and observational techniques. Here are several key aspects of the role of astrophysics in understanding the universe:

Unraveling Cosmos's Origins and Evolution: By studying the cosmic microwave background radiation, the structure and distribution of galaxies, and the cosmic web, astrophysicists seek to unravel the origins and evolution of the universe. This includes exploring the nature of dark matter, dark energy, and the cosmic inflation that shaped the early universe.

Investigating Stellar Life Cycles: Astrophysics provides insights into the life cycles of stars, from their formation in nurseries of gas and dust to their evolution into supernovae, neutron stars, or black holes. This understanding includes the nuclear processes that fuel stars and the mechanisms governing their eventual demise.

Exploring Exoplanets and Planetary Systems: The study of exoplanets, planetary systems, and their host stars deepens our knowledge of the diversity of worlds beyond our solar system. Astrophysical techniques, such as transit photometry and radial velocity measurements, enable the discovery and characterization of exoplanets, informing our understanding of planetary formation and potential habitability.

Shedding Light on Galactic Dynamics: Astrophysics delves into the dynamics of galaxies, their formation, interactions, and the nature of supermassive black holes at their centers. This encompasses the study of galactic morphology, the distribution of dark matter, and the coevolution of galaxies and their central black holes.

Probing Fundamental Physics: From testing the principles of general relativity in extreme astrophysical environments to investigating high-energy astrophysical phenomena, astrophysicists contribute to our understanding of fundamental physics on cosmic scales. This includes the study of gravitational waves, cosmic rays, and high-energy particle astrophysics.

Exploring Cosmic Phenomena and Cosmology: The field of astrophysics encompasses the study of various cosmic phenomena, including pulsars, quasars, gamma-ray bursts, and active galactic nuclei. The insights gained contribute to our understanding of cosmic structure, the nature of spacetime, and the evolution of the universe on large scales.

Advancing Observational and Computational Techniques: Astrophysics drives advancements in observational techniques, instrumentation, and computational modeling, enabling the study of the universe across multiple electromagnetic wavelengths and simulation of complex astrophysical processes.

By encompassing these diverse aspects, astrophysics serves as a foundational discipline in our quest to comprehend the cosmos, continually expanding the frontiers of knowledge, inspiring wonder, and deepening our appreciation of the vast, dynamic, and awe-inspiring universe in which we dwell.

Technological advancements aiding alien life search

Technological advancements have greatly enhanced humanity's ability to search for alien life, enabling scientists to explore the cosmos with unprecedented precision and depth. Here are some key technological developments that have facilitated the search for extraterrestrial life:

Exoplanet Detection and Characterization: Advanced telescopes and instruments, such as the Kepler Space Telescope and the Transiting Exoplanet Survey Satellite (TESS), have revolutionized the detection of exoplanets. High-precision photometry and spectroscopy enable the identification and characterization of exoplanets, including their sizes, orbits, and atmospheric compositions.

Spectroscopic Analysis: Cutting-edge spectroscopic techniques, including high-resolution spectroscopy and transit spectroscopy, allow scientists to analyze the chemical composition of exoplanet atmospheres. This includes the detection of potential biosignatures, such as the presence of water vapor, oxygen, and methane, that may indicate the presence of life.

Adaptive Optics and Interferometry: Adaptive optics and interferometric techniques enhance the resolution and sensitivity of telescopes, enabling high-precision imaging and spectroscopy of exoplanetary systems. These technologies provide detailed insights into the architecture and dynamics of exoplanetary systems, including the potential for habitable environments.

Space-based Observatories: Space-based observatories, such as the Hubble Space Telescope and the James Webb Space Telescope, offer unobstructed views of the cosmos and access to ultraviolet, optical, and infrared wavelengths. These platforms enable the study of exoplanetary atmospheres, the search for planetary disks, and the exploration of potential habitats for extraterrestrial life.

Advanced Data Processing and Analysis: State-of-the-art data processing and analysis techniques, including machine learning algorithms and computational modeling, aid in the interpretation of complex astrophysical data. These tools enable the identification of exoplanets, the analysis of planetary atmospheres, and the extraction of valuable insights from large datasets.

Robotic Probes and Landers: Robotic probes and landers, such as the Mars rovers and upcoming missions to icy moons, facilitate the exploration of potential extraterrestrial habitats within our own solar system. These technologies enable the study of planetary environments and the search for signs of past or present life beyond Earth.

Radio and Optical SETI: The Search for Extraterrestrial Intelligence (SETI) benefits from advanced radio and optical telescopes, as well as sophisticated signal processing algorithms. These tools enhance the sensitivity and scope of SETI initiatives, enabling the detection and analysis of potential signals from extraterrestrial civilizations.

Next-Generation Space Missions: Future space missions, including the James Webb Space Telescope, the LUVOIR concept, and the European Space Agency's PLATO mission, promise to revolutionize our understanding of exoplanets and potentially habitable worlds. These endeavors will leverage advanced technologies to expand the frontiers of cosmic exploration.

These technological advancements collectively contribute to the ongoing search for alien life, empowering scientists to push the boundaries of knowledge and capability as they endeavor to uncover the secrets of life beyond Earth.

Key discoveries of exoplanets and the potential they hold

The study of exoplanets has yielded key discoveries that have transformed our understanding of planetary systems and the potential for habitable worlds beyond our solar system. Some of the key discoveries and the potential they hold include:

First Exoplanet Detection (1992): The discovery of the first exoplanet orbiting a main-sequence star, PSR B1257+12, marked a historic breakthrough, demonstrating that planets exist beyond our solar system. This discovery ignited a new era of exoplanet research and exploration.

Exoplanet Diversity and Abundance: The detection of thousands of exoplanets of diverse sizes, compositions, and orbital characteristics has revealed the rich abundance and diversity of planetary systems in our galaxy. This diversity includes gas giants, terrestrial planets, super-Earths, and exoplanets within habitable zones.

Habitable Zone Exoplanets: The identification of exoplanets within the habitable zones of their host stars—regions where conditions may be suitable for liquid water—has raised the tantalizing prospect of habitable worlds beyond Earth. These discoveries expand the range of celestial bodies that may harbor life as we know it.

Exoplanetary Atmospheres and Biosignatures: The analysis of exoplanetary atmospheres has offered insights into their compositions, including the detection of water vapor, carbon dioxide, methane, and other chemicals. The identification of potential biosignatures within exoplanetary atmospheres has bolstered the prospects of detecting signs of life beyond Earth.

Exoplanet Transits and Atmospheric Characterization: The observation of exoplanetary transits, where a planet crosses in front of its host star, has enabled scientists to study exoplanetary atmospheres using techniques such as transmission spectroscopy. This approach provides valuable data on atmospheric compositions and characteristics.

Exoplanetary Systems and Dynamics: Studies of exoplanetary systems have revealed complex architectures, including multiple-planet systems, resonant orbits, and interactions between planets and their host stars. These findings inform our understanding of planetary formation, migration, and stability within planetary systems.

Search for Earth-like Exoplanets: Ongoing efforts to identify Earth-like exoplanets, rocky worlds with potential for liquid water and habitability, hold promise for finding planetary analogs to our own world. Such discoveries could offer comparative insights into the conditions conducive to life.

Rarity of Unusual Systems: The detection of rare or unusual exoplanetary systems, such as hot Jupiters, rogue planets, and circumbinary planets orbiting binary stars, expands our knowledge of the diverse and sometimes unexpected configurations of planetary systems.

These key discoveries collectively underscore the potential for a multitude of planetary environments and the diverse pathways to habitability in the cosmos, fueling the ongoing quest to unravel the cosmic enigma of life beyond Earth.

The Drake Equation as a tool for estimating life

The Drake Equation is a conceptual framework used to estimate the potential number of extraterrestrial civilizations in our galaxy that could be capable of communicating with us. Formulated by Dr. Frank Drake in 1961, the equation synthesizes various factors to provide a probabilistic estimate of the existence of communicative extraterrestrial civilizations. The equation is as follows:

$$ N = R_* \times f_p \times n_e \times f_l \times f_i \times f_c \times L $$

Here's a breakdown of the components of the equation:

(N) = The number of civilizations in our galaxy that we could potentially communicate with.

(R_*) = The average rate of star formation in our galaxy.

(f_p) = The fraction of those stars that have planets.

(n_e) = The average number of planets that could potentially support life per star that has planets.

(f_l) = The fraction of planets that could support life that actually develop life at some point.

(f_i) = The fraction of planets with life that actually go on to develop intelligent life.

(f_c) = The fraction of planets with intelligent life that are capable of communicating signals into space.

(L) = The length of time such civilizations release detectable signals into space.

The Drake Equation is a tool for stimulating scientific dialogue and inquiry, guiding discussions on the potential factors that could contribute to the existence of communicative extraterrestrial civilizations. However, it is important to note that due to uncertainties and limited knowledge about the values of the variables in the equation, the estimate derived from the Drake Equation is highly speculative and subject to wide variation.

It should be emphasized that the Drake Equation serves primarily as a conceptual framework for considering the factors that may contribute to the existence of communicative extraterrestrial civilizations, rather than as a precise predictive tool. As our understanding of exoplanets, astrobiology, and the conditions necessary for life continues to advance, the Drake Equation remains a thought-provoking construct that fosters contemplation and exploration of the potential for life in the universe.

The Dynamics of Life: Understanding Habitability

Understanding the dynamics of habitability is crucial for evaluating the potential for life beyond Earth and exploring the conditions that may support living organisms in diverse planetary environments. This concept encompasses a wide range of interdisciplinary considerations, including astrobiology, planetary science, and environmental conditions. Here are key aspects of the dynamics of habitability:

Planetary Conditions: Habitability is fundamentally tied to the environmental conditions of a planetary body, including its atmosphere, surface composition, and geophysical processes. Understanding factors such as temperature, pressure, radiation levels, and the presence of liquid water is essential for assessing the potential habitability of a celestial body.

Astrobiological Factors: Astrobiology explores the fundamental principles that govern the origins and sustenance of life, including the search for biosignatures, the study of extremophiles on Earth, and the identification of potential habitats beyond our planet. This interdisciplinary field contributes to our understanding of the conditions conducive to life in diverse environments.

Habitability Zones: Identifying habitable zones around stars, where conditions may be suitable for the existence of liquid water and potentially life-sustaining environments, is a key aspect of understanding habitability. This involves considering the orbital distance from a star where temperatures could support the presence of liquid water on a planetary surface.

Planetary Atmospheres: The study of planetary atmospheres, including their compositions, interactions with stellar radiation, and the potential presence of greenhouse gases, is critical for assessing habitability. Understanding the dynamics of atmospheric processes and the potential for stable, life-friendly conditions contributes to our understanding of planetary habitability.

Planetary Interiors and Tectonics: Exploring the interior dynamics of planetary bodies, including their geological activity, magnetic fields, and potential for subsurface environments, informs our understanding of habitability. The presence of tectonic processes, geothermal energy, and the protection of planetary interiors contribute to the potential for habitable conditions.

Stellar and Galactic Influences: Considering the influence of stellar and galactic factors, such as stellar radiation, cosmic rays, and the interstellar environment, on the habitability of planetary systems is essential. Understanding how celestial phenomena and galactic dynamics impact the potential for life expands our knowledge of habitability.

Evolution of Complex Life: The dynamics of habitability also encompass the potential for the evolution of complex, multicellular life forms on habitable worlds. Understanding the factors that contribute to the emergence and sustenance of complex life is a key dimension of habitability studies.

By integrating these diverse considerations, scientists seek to unravel the dynamics of habitability in a variety of planetary environments, fostering a holistic understanding of the conditions that may support life beyond Earth and informing the ongoing search for habitable worlds and potential extraterrestrial life.

Defining habitability in terms of extraterrestrial life

Habitability, in the context of extraterrestrial life, refers to the suitability of a planetary environment to support the emergence, sustenance, and potential evolution of living organisms. It encompasses a wide array of physical, chemical, and environmental conditions that are conducive to the development of life as we know it, as well as the exploration of possible alternative forms of life.

Key considerations in defining the habitability of a planetary environment for extraterrestrial life include:

Presence of Liquid Water: Liquid water is a fundamental requirement for life as we know it. Habitability often involves assessing the potential for liquid water to exist on or below the surface of a planetary body, as well as its availability over geological timescales.

Stable and Suitable Temperatures: Habitability considers the range of temperatures that allows for the presence of liquid water and the conditions necessary for the chemical reactions underlying life processes. This involves assessing the stability of temperatures within a range that is compatible with biological activity.

Chemical Composition and Nutrient Availability: Understanding the chemical composition of a planetary environment, including the presence of essential elements and organic molecules, is vital for evaluating habitability. This involves assessing the potential for nutrient availability and the presence of chemical energy sources that could support life.

Atmospheric Conditions: The study of atmospheric properties, including the presence of oxygen, carbon dioxide, and other gases, informs our understanding of habitability. This includes considering the potential greenhouse effect, the shielding of harmful radiation, and the interaction between the atmosphere and the planetary surface.

Stability and Protection from Hazards: Habitability involves assessing the stability of planetary environments and the protection of potential habitats from hazards such as extreme geological events, high levels of radiation, and cosmic impact events.

Planetary Interiors and Geophysical Factors: Understanding the geophysical processes of a planetary body, including tectonics, volcanic activity, and the potential for subsurface environments, contributes to the assessment of habitability.

Potential for Evolution and Sustenance of Life: Habitability encompasses an exploration of the potential for life to emerge, evolve, and persist within a given environment, including the presence of energy sources, the availability of habitable niches, and the potential for biological diversity.

By considering these diverse factors, scientists work to define the habitability of exoplanets, planetary moons, and other celestial bodies in the ongoing quest to understand the conditions that may support extraterrestrial life, as well as the potential for alternative forms of life in the cosmos.

The Goldilocks Zone and its significance

The "Goldilocks Zone," also known as the "habitable zone," refers to the region around a star where conditions are conducive to the existence of liquid water on the surface of a planetary body. This concept is crucial in the search for potential habitable exoplanets and the assessment of their suitability to support life. The significance of the Goldilocks Zone lies in several key aspects:

Potential for Liquid Water: The Goldilocks Zone represents the range of orbital distances from a star where the temperature is neither too hot nor too cold for liquid water to exist on the surface of a planetary body. Given that liquid water is a fundamental requirement for life as we know it, the presence of this zone informs our search for habitable worlds beyond our solar system.

Habitability Assessment: Identifying exoplanets within the habitable zone provides a starting point for assessing their potential suitability to support life. While the presence of liquid water is a critical factor, habitability involves the consideration of additional environmental conditions, such as atmospheric composition, stability, and the presence of essential chemical elements.

Targeting Observations and Studies: The existence of the Goldilocks Zone guides astronomers and exoplanet researchers in targeting their observations and studies. The quest to identify and characterize exoplanets within this zone, using techniques such as transit photometry, radial velocity measurements, and direct imaging, is aimed at uncovering potentially habitable environments.

Candidates for Biosignatures: Exoplanets within the habitable zone are considered prime candidates for the search for biosignatures—indicators of the presence of life or conditions conducive to life, such as the detection of oxygen, methane, and other chemical markers that may point to the possibility of biological activity.

Expanding the Search for Life: The presence of the Goldilocks Zone broadens the scope of our search for life beyond Earth. By focusing on exoplanets within this zone, researchers aim to explore a wider range of potential habitats, expanding the frontiers of our understanding of the diversity and potential for life in the universe.

Overall, the concept of the Goldilocks Zone is significant in guiding the search for potentially habitable exoplanets, informing our understanding of the conditions that may support life, and fostering the ongoing exploration of the cosmic enigma of extraterrestrial habitability.

Astrobiological factors critical for life

Several astrobiological factors are critical for the emergence, sustenance, and potential evolution of life beyond Earth. These factors encompass a wide range of interrelated conditions and processes that are essential for the development and support of living organisms in diverse planetary environments. Some key astrobiological factors critical for life include:

Presence of Liquid Water: Liquid water is recognized as a fundamental requirement for life as we know it. Its unique properties as a solvent and medium for biochemical reactions make it essential for the emergence and sustenance of biological processes.

Chemical Composition and Essential Elements: The availability of essential chemical elements, such as carbon, hydrogen, oxygen, nitrogen, phosphorus, and sulfur, is critical for the formation of organic molecules and the building blocks of life. The presence of these elements in planetary environments provides the fundamental components necessary for biological activity.

Energy Sources: The availability of energy sources, in the form of chemical energy, light, geothermal heat, or other sources, is vital for driving biological processes and sustaining life. Chemical gradients and energy sources support metabolic activities and the potential sustenance of living organisms.

Stable Planetary Environments: Habitability involves assessing the stability of planetary environments, including the presence of stable climate conditions, suitable temperature ranges, and protection from extreme events such as cosmic impacts, stellar flares, or geological disturbances.

Atmospheric Composition and Greenhouse Gases: The composition of planetary atmospheres, including the presence of greenhouse gases, influences the retention of heat, the regulation of surface temperatures, and the potential for liquid water to exist on a planetary body's surface.

Planetary Dynamics and Tectonics: Understanding the geophysical processes of planetary bodies, including tectonic activity, geological cycles, and the potential for geothermal energy, contributes to the assessment of habitability and the availability of stable environments.

Planetary Habitats and Niches: Exploring the presence of potential habitats and ecological niches—such as subsurface environments, hydrothermal vents, or regions shielded from harmful radiation—contributes to our understanding of the diverse conditions that may support life in different planetary environments.

Biosignatures and Biological Markers: The search for biosignatures, such as the detection of gases like oxygen, methane, and other chemical markers, provides insight into the potential presence of life or conditions conducive to life on a planetary body.

By considering these interdependent factors, astrobiologists and researchers endeavor to unravel the complex dynamics that underpin the potential for life beyond Earth, fostering a holistic understanding of the astrobiological conditions that may offer suitable environments for the emergence and sustenance of living organisms in the cosmos.

Limits of life on Earth as a blueprint

The limits of life on Earth serve as a valuable blueprint for understanding the potential boundaries and adaptability of life in diverse environments, as well as informing the search for extraterrestrial life. By studying the extremophilic organisms and the various environments where life thrives on Earth, scientists gain insights into the adaptability and resilience of life in extreme conditions. Here are some key aspects of how the limits of life on Earth serve as a blueprint:

Extreme Environmental Adaptation: Extremophiles, microorganisms that thrive in extreme environments such as high temperatures, high-pressure conditions, acidic or alkaline environments, and high radiation areas, provide a window into the adaptability of life. Understanding the strategies and biochemical adaptations of extremophiles informs our knowledge of the potential for life to exist in extraterrestrial environments with similar extreme conditions.

Adaptation to Low Nutrient Environments: Organisms that survive and thrive in low-nutrient environments, such as deep ocean hydrothermal vents or subsurface habitats, offer insights into the minimal requirements for life and the potential for metabolic adaptability in resource-limited settings. This information informs the search for life in similar environments beyond Earth.

Radiation Tolerance and Space Conditions: Studies on the radiation tolerance of extremophiles and their survival in space-like conditions contribute to our understanding of the potential for life to endure in extraterrestrial environments, such as the surface of Mars, icy moons, or interplanetary space.

Biogeological Processes: The study of extremophiles in the context of biogeological processes, such as mineralization, geochemical cycling, and energy metabolism, offers insights into the interconnectedness of life with planetary environments and the potential for life to influence and be influenced by the geology and chemistry of its surroundings.

Considerations for Habitability: The adaptation and survival strategies of extremophiles inform our understanding of habitability considerations for extraterrestrial environments, guiding the search for potential habitats and the assessment of the limits and constraints for supporting life beyond Earth.

By drawing from the blueprint of Earth's extremophiles and their adaptations to diverse and extreme conditions, scientists are better equipped to assess the potential for life in a variety of extraterrestrial environments, inform the search for habitable worlds beyond our solar system, and expand our understanding of life's potential adaptability in the cosmos.

Unraveling Messages from the Cosmos: SETI and Beyond

The search for extraterrestrial intelligence (SETI) and the broader quest to unravel potential messages from the cosmos represent a fascinating and multidisciplinary endeavor involving astronomy, astrophysics, signal processing, and technological innovation. Beyond the traditional scope of astronomical research, SETI and related initiatives delve into the prospect of detecting signals from potential extraterrestrial civilizations and exploring the implications of such a discovery.

Radio and Optical Searches: SETI encompasses efforts to detect potential signals from extraterrestrial civilizations, including targeted radio and optical searches for anomalous, non-natural signals that may indicate artificial origin. These endeavors involve scanning the cosmos for narrowband signals, broadband transmissions, or pulsed emissions that stand out from background noise.

Advancements in Signal Processing: The field of SETI is augmented by advancements in signal processing, data analysis techniques, and machine learning algorithms that aid in the identification and classification of potential extraterrestrial signals. These tools bolster the sensitivity and efficiency of SETI searches.

Technological Innovation: SETI initiatives benefit from ongoing technological innovation, including the development of increasingly sensitive radio telescopes, optical observatories, and data processing systems capable of sifting through vast amounts of cosmic data for potential signals.

Astrophysical Constraints and Target Selection: The search for extraterrestrial signals involves considering astrophysical constraints, such as the properties of target stars, the suitability of planetary systems, and the selection of stellar environments with the potential for hosting extraterrestrial civilizations.

Communication Integrity and Ethics: Deliberations on the integrity of interstellar communication, the ethical implications of potential contact with extraterrestrial civilizations, and the principles guiding responsible interstellar messaging contribute to the ethical dimensions of SETI and the considerations surrounding potential discoveries.

Interdisciplinary Collaboration: The field of SETI fosters interdisciplinary collaboration, drawing expertise from astronomy, astrophysics, planetary science, anthropology, linguistics, and ethics to inform the multifaceted dimensions of the search for extraterrestrial signals.

Beyond Traditional SETI Approaches: Recent efforts in SETI have expanded to include novel approaches, such as the study of interstellar objects, transient phenomena, and the search for potential artifacts or techno signatures that may point to the existence of extraterrestrial intelligence.

The quest to unravel messages from the cosmos through SETI and related endeavors continues to inspire scientific inquiry, technological innovation, and contemplation of humanity's place in the cosmos. While the detection of extraterrestrial signals remains an ongoing scientific pursuit with profound implications, the broader impact of SETI extends to igniting public interest in science, stimulating philosophical reflection, and fostering a sense of wonder and curiosity about the potential for contact with other intelligent beings in the universe.

The mission and methods of the Search for Extraterrestrial Intelligence (SETI)

The Search for Extraterrestrial Intelligence (SETI) is a scientific endeavor dedicated to the exploration of the cosmos for potential signals indicative of extraterrestrial civilizations. This concerted effort involves the use of specialized instruments, innovative methods, and multidisciplinary approaches to detect anomalous signals that may point to the existence of intelligent life beyond Earth. Here are the mission and methods of the Search for Extraterrestrial Intelligence (SETI):

Mission: The primary mission of SETI is to search for evidence of extraterrestrial intelligence by detecting and interpreting signals or phenomena that could be of artificial origin. This includes observing potential electromagnetic transmissions across radio and optical wavelengths, as well as considering the possibilities of interstellar communication, technological artifacts, or deliberate attempts at contact from advanced civilizations.

Methods: The methods employed in the search for extraterrestrial intelligence encompass a range of observational, technological, and analytical techniques aimed at detecting potential signals from space. These methods include:

Radio Searches: Radio telescopes are used to scan the sky for narrowband or broadband radio signals that stand out from natural cosmic sources, such as pulsars, quasars, and astrophysical phenomena. This involves targeted searches of specific star systems and broad surveys of the sky to identify potential anomalous emissions.

Optical Searches: Optical observatories conduct surveys of star systems and planetary environments to look for potential optical signals, such as laser emissions or unusual patterns of light that may indicate artificial origin.

Signal Processing and Data Analysis: Advanced signal processing techniques and data analysis algorithms are employed to sift through massive datasets and identify potential candidate signals. This includes the use of machine learning, statistical analysis, and multi-spectral processing to differentiate natural signals from potential artificial emissions.

Technological Innovation: SETI initiatives leverage ongoing technological advances in radio astronomy, optical spectroscopy, data processing, and telescopic instrumentation to enhance the sensitivity and precision of signal detection efforts.

Target Selection and Astrophysical Constraints: SETI researchers consider astrophysical constraints, the properties of target stars, the potential habitability of exoplanetary systems, and the suitability of planetary environments when selecting regions for signal detection and analysis.

Interdisciplinary Collaboration: SETI fosters collaboration across diverse scientific disciplines, drawing expertise from astronomy, astrophysics, engineering, computer science, and data analysis to inform the design, implementation, and interpretation of signal detection methods.

By pursuing these methods, SETI endeavors to explore the cosmos for potential signals from extraterrestrial civilizations, stimulating scientific inquiry, technological innovation, and contemplation of our place in the universe. While the quest for evidence of extraterrestrial intelligence remains ongoing, the commitment to rigor and methodical exploration bolsters the profound scientific and philosophical implications of SETI initiatives.

Analysis of radio signals and what they entail

The analysis of radio signals in the context of the search for extraterrestrial intelligence (SETI) involves a multifaceted and rigorous approach aimed at discerning potentially anomalous, artificial emissions from the natural backdrop of cosmic radio sources. Here's an overview of the analysis of radio signals in SETI and what they entail:

Radio Signal Collection: Radio telescopes and observatories are employed to capture and record radio waves across a broad range of frequencies. These signals originate from celestial sources, encompassing natural emissions from stars, pulsars, galaxies, and astrophysical phenomena, as well as potential artificial signals that may be of interest in the context of SETI.

Spectrum Analysis: Analysis of the radio spectrum involves examining the distribution of signal power across different radio frequency bands. Researchers scrutinize the power spectra of recorded signals to determine their frequency, bandwidth, and potential modulation patterns that deviate from typical natural emissions.

Detection of Narrowband Signals: SETI efforts seek to identify narrowband signals that exhibit an unusually concentrated energy at specific frequencies, which may be indicative of deliberate, targeted transmissions from extraterrestrial civilizations. Analyses involve discerning narrow spectrally confined emissions that differ from the broader spectral signatures of natural cosmic sources.

Search for Pulsed or Modulated Signals: The quest for pulsed or modulated radio signals is a key aspect of SETI analysis. Pulsed emissions, or signals with periodic or rhythmic characteristics, as well as modulated transmissions that exhibit variations in amplitude, frequency, or phase, are subject to detailed scrutiny for potential artificial origins.

Interference Mitigation: Analysis efforts aim to mitigate and account for terrestrial and human-generated radio frequency interference, such as radio broadcasts, satellite communications, and other sources of anthropogenic emissions that may obscure or mimic potential extraterrestrial signals.

Machine Learning and Pattern Recognition: Advanced algorithms, including machine learning, statistical pattern recognition, and computational methods, are employed to aid in the classification and identification of candidate signals. These tools assist in distinguishing potential artificial emissions from natural cosmic radio sources.

Follow-up Observations and Confirmation: Candidate signals identified through initial analyses may undergo further scrutiny through follow-up observations, observations of the same source from multiple observatories, and complementary data collection to confirm the consistency and persistence of potential signals.

The analysis of radio signals in the context of SETI entails a meticulous, methodical, and rigorous examination of cosmic radio emissions, aimed at discerning potential anomalies that may signify the presence of artificial signals of extraterrestrial origin. This comprehensive approach, leveraging technological innovation, interdisciplinary collaboration, and meticulous scrutiny, informs the ongoing quest to explore the cosmos for evidence of extraterrestrial intelligence.

Future prospects in identifying alien communications

Future prospects in identifying alien communications are shaped by ongoing advancements in observational techniques, data analysis, and technological innovation. These prospects are poised to address several key areas of development in the search for extraterrestrial intelligence (SETI) and the endeavor to decipher potential messages from the cosmos. Here are some future prospects in the identification of alien communications:

Advancements in Radio and Optical Observations: Future developments in radio and optical observatories, including the construction of next-generation telescopes and specialized instrumentation, will bolster the sensitivity, range, and precision of signal detection efforts. These advancements will expand the reach of SETI searches and enhance the capability to identify potential alien communications.

Next-Generation Data Processing and Analysis: Ongoing progress in signal processing, machine learning, and computational algorithms will further refine the capacity to discern potential anomalies in cosmic data and facilitate the identification of candidate signals from extraterrestrial sources. Future developments in data analysis techniques will aid in the swift, automated assessment of vast quantities of observational data.

Interferometer Arrays and Multiwavelength Studies: The deployment of interferometer arrays and multiwavelength observational campaigns will enable comprehensive studies of potential candidate signals, offering complementary information across a range of radio and optical frequencies. These multiwavelength studies will provide new insights into the properties and origins of anomalous emissions.

Targeted Exoplanet Searches: Future SETI efforts will focus on targeted searches of exoplanetary systems, taking advantage of the identification and characterization of habitable exoplanets. These targeted studies will prioritize star systems with the potential for hosting habitable worlds and aim to discern potential signals from these promising targets.

Global Collaborative Initiatives: Enhanced international collaboration, resource sharing, and coordination among SETI researchers and observatories will foster a unified, global effort in the search for extraterrestrial intelligence. Interdisciplinary partnerships and collaborative initiatives will leverage diverse expertise and resources to propel the search for alien communications.

Technological Innovation and Instrumentation: Ongoing advancements in technology, including developments in phased array radio antennas, next-generation spectrographs, and space-based observatories, will drive the expansion and diversification of observational capabilities, opening new avenues for identifying potential alien communications.

Interstellar Messaging and Active SETI Initiatives: Considerations for interstellar messaging and deliberated attempts to broadcast intentional signals to specific target stars will contribute to the broader dialogue on potential methods for initiating interstellar communication. Future prospects may involve the pursuit of responsible and ethical practices in active SETI initiatives.

Patterns of Technosignatures and Artifacts: SETI prospects encompass the study of potential technosignatures, artifacts, or unconventional phenomena that may hint at the presence of advanced extraterrestrial technologies, structures, or emissions within our cosmic surroundings.

By aligning with these future prospects, SETI and related initiatives are positioned to advance the frontiers of cosmic exploration, deepen our understanding of the potential for extraterrestrial intelligence, and engage in ongoing efforts to unravel potential messages from the cosmos.

The debate around sending messages into space

The debate around sending messages into space, particularly intentional interstellar communication, reflects a multifaceted and thought-provoking discourse that spans scientific, ethical, sociocultural, and philosophical considerations. The prospect of deliberate transmissions from Earth to potential extraterrestrial civilizations, known as active SETI or messaging to extraterrestrial intelligence (METI), engenders a range of perspectives and deliberations. Here are key facets of the debate:

Scientific Inquiry and Interest: Proponents of active SETI emphasize the potential scientific value of intentionally sending messages into space, arguing that such efforts can stimulate interstellar communication, foster global collaboration, and provoke discussion on humanity's place in the cosmos.

Interstellar Communication Paradigms: The debate intersects with differing paradigms regarding the desirability and feasibility of interstellar communication. Advocates of active SETI contend that initiating intentional transmissions may form part of an interstellar conversation, while skeptics raise concerns about the motivations, risks, and potential implications of such initiatives.

Ethical Considerations: The ethical dimensions of active SETI initiatives are a central focus of the debate. Deliberations encompass the potential consequences of interstellar messaging, the unforeseeable impact on terrestrial and extraterrestrial societies, and the ethical responsibility associated with representing humanity in the context of cosmic communication.

Risk Assessment and Precautionary Principles: Skeptics of active SETI underscore the importance of exercising precautionary principles and risk assessment in interstellar messaging, considering the uncertainties and potential ramifications of intentionally broadcasting signals into space.

Cultural and Societal Representations: Sociocultural dimensions emerge in the debate, encompassing discussions about humanity's representation in potential interstellar communications, the portrayal of terrestrial cultures, and the diversity of values and perspectives that inform the desirability of initiating contact with extraterrestrial civilizations.

Global Governance and Consultative Processes: The dialogue extends to considerations of international governance, consultation, and global deliberation in the context of METI initiatives. Engagement with regulatory frameworks, coordination among nations, and public engagement in active SETI efforts contribute to the broader debate.

Long-Term Implications of Contact: Contemplation of the potential long-term implications of interstellar communication, including the impact on terrestrial societies, the potential for a response from extraterrestrial civilizations, and the far-reaching consequences of initiating contact, inform the depth and complexity of the debate.

Alternatives and Passive SETI: The debate around active SETI intersects with discussions about alternative methods of interstellar communication and the pursuit of passive SETI efforts, which focus on listening for potential signals from space rather than transmitting intentional messages.

Overall, the debate around sending intentional messages into space reflects a rich tapestry of scientific, ethical, cultural, and philosophical considerations, inviting ongoing dialogue, insight, and reflection on the potential impact of interstellar communication on humanity and our place in the cosmos.

Evaluating Evidence: UFOs, Encounters, and Science

The evaluation of evidence related to unidentified flying objects (UFOs), encounters, and alleged extraterrestrial phenomena involves a rigorous and critical assessment within the framework of scientific inquiry and empirical evidence. While UFO sightings and encounters have generated widespread interest and speculation, the scientific community approaches these claims with a commitment to evidence-based analysis and skepticism. Here are key aspects of evaluating evidence related to UFOs, encounters, and the scientific perspective:

Observational Data and Testimonial Evidence:
Evaluating reports of UFO sightings and encounters

involves scrutinizing observational data, eyewitness accounts, and testimonial evidence. This entails assessing the credibility, consistency, and corroboration of witness testimonies, as well as the quality of documented observations.

Extraterrestrial Hypothesis and Alternative Explanations: Scientific evaluation involves considering the extraterrestrial hypothesis as well as alternative explanations for UFO sightings, encompassing natural atmospheric phenomena, astronomical objects, weather-related occurrences, and human-made aerial activities. This process aims to rule out conventional explanations before considering extraordinary claims.

Empirical Analysis and Physical Evidence: Rigorous examination of physical evidence, including photographic documentation, radar data, sensor recordings, and material samples, contributes to the scientific evaluation of alleged UFO encounters. This analysis involves authenticating and validating evidence to substantiate extraordinary claims.

Psychosocial Factors and Human Perception: Consideration of psychosocial factors, cognitive biases, human perception, and psychological perceptions plays a role in assessing the reliability and interpretational aspects of UFO encounters. This encompasses the study of eyewitness testimony, memory recall, and perceptual phenomena.

Scientific Method and Hypothesis Testing: The scientific community applies the principles of hypothesis testing, empirical investigation, and falsifiability in the assessment of UFO claims. This methodical approach involves formulating testable hypotheses, conducting controlled experiments, and seeking reproducible evidence to support extraordinary assertions.

Peer Review and Scholarly Scrutiny: Elevating UFO claims to scientific scrutiny involves subjecting evidence to peer review within the scholarly community. This process entails critical analysis, validation of methodology, and transparent discourse among experts in relevant fields.

Misidentifications and Anomalies: Recognition of the potential for misidentifications, perceptual illusions, misinterpretations, and anomalous visual phenomena informs the comprehensive evaluation of UFO claims. This includes consideration of cultural influences, context-dependent factors, and reporting biases in observational data.

Public Policy and Regulatory Oversight: The evaluation of UFO evidence intersects with considerations of public policy, regulatory oversight, and national security dimensions. Coordination among governmental agencies, scientific bodies, and public engagement informs the responsible management of potential safety, security, and societal implications.

By applying these principles, the scientific community seeks to maintain a balanced, evidence-based perspective on UFO encounters, extraterrestrial claims, and the pursuit of empirical understanding while addressing public interest and discourse surrounding these phenomena.

Historical accounts of Unidentified Flying Objects (UFOs)

Historical accounts of Unidentified Flying Objects (UFOs) date back centuries and encompass a wide array of documented sightings, encounters, and reports of aerial phenomena that defy conventional explanation. These historical accounts contribute to the enduring intrigue and speculation surrounding UFOs, inspiring scientific inquiry, public interest, and cultural fascination. Here are some notable historical accounts of UFOs:

Early Historical Records: Throughout history, there are records of anomalous aerial phenomena, including ancient writings, medieval chronicles, and historical manuscripts that describe celestial apparitions, mysterious lights, and unexplained aerial objects.

Early Modern Era: Accounts of unusual aerial sightings gained prominence during the early modern era and the age of exploration, with reports of enigmatic lights, celestial disturbances, and unusual airborne phenomena that captured the attention of astronomers, natural philosophers, and scholars of the time.

Foo Fighters and Ghost Rockets: During World War II and the postwar period, aviators reported sightings of enigmatic aerial objects, referred to as "foo fighters" by Allied pilots and "ghost rockets" by observers in Europe. These sightings presented perplexing phenomena that eluded easy identification.

Emergence of UFO Terminology: The term "flying saucer" gained widespread usage following the highly publicized 1947 report by pilot Kenneth Arnold, who described seeing peculiar aircraft-like objects near Mount Rainier, Washington. This event catalyzed global interest in UFOs and popularized the term "flying saucer."

Roswell Incident: The 1947 Roswell incident, an alleged UFO crash in New Mexico, became a focal point of UFO lore and conspiracy theories, contributing to the enduring mythos and cultural impact surrounding alleged extraterrestrial visitation.

Official Investigations and Reports: Governmental investigations and official reports of UFO sightings, including the U.S. Air Force's Project Blue Book, the Condon Committee, and other initiatives, documented and examined thousands of reports, seeking to discern patterns and ascertain the nature of unexplained aerial phenomena.

Global Sightings and Cultural Impact: UFO sightings and reports have been documented across the globe, with widespread accounts from diverse cultures, regions, and historical periods. These sightings have influenced cultural beliefs, artistic expressions, and sociopolitical discourse relating to the potential existence of extraterrestrial visitation.

Media Coverage and Popular Culture: UFO sightings have garnered extensive media coverage, with numerous reports, documentaries, books, and films exploring the phenomenon. This exposure has shaped public fascination and contributed to enduring interest in UFOs.

Historical accounts of UFOs offer a rich tapestry of documented sightings, intriguing anomalies, and cultural significance, informing ongoing interest, scientific inquiries, and societal dialogues surrounding unexplained aerial phenomena. These historical records continue to inspire diverse perspectives, scientific analyses, and public engagement with the enigmatic realm of UFO encounters.

Scientific scrutiny of alleged encounters

The scientific scrutiny of alleged encounters with unidentified flying objects (UFOs) involves a methodical and evidence-based approach aimed at critically evaluating the reported incidents within the framework of empirical analysis, skepticism, and rigorous inquiry. This process aims to discern the nature of reported phenomena, consider natural or conventional explanations, and discern the potential validity or falsifiability of UFO claims. Here are key aspects of the scientific scrutiny of alleged UFO encounters:

Empirical Analysis of Observational Data: Examination of observational data, including eyewitness accounts, photographs, videos, and radar tracking, involves scrutinizing the reliability, consistency, and corroborative nature of reported incidents. This analysis forms the basis for assessing the credibility and evidentiary quality of alleged encounters.

Natural and Conventional Explanations: Scientific scrutiny involves considering natural atmospheric phenomena, astronomical objects, human-made aircraft, drones, or other terrestrial sources as potential explanations for observed UFO incidents. This process aims to rule out conventional causes before considering extraordinary claims.

Psychosocial and Perceptional Factors: Evaluation of psychosocial factors, such as cognitive biases, misinterpretations, perceptual illusions, and social influences, informs the assessment of human perception in alleged UFO encounters. This consideration illuminates the potential for misidentifications and misconceptions in reported incidents.

Physical and Forensic Analysis: Rigorous examination of physical evidence, including materials, traces, and environmental effects allegedly associated with UFO incidents, contributes to the scientific scrutiny of claimed encounters. This analysis involves authenticating evidence and discerning any physical anomalies.

Hypothesis Testing and Critical Assessment:
Application of scientific principles, including hypothesis testing, falsifiability, and critical evaluation, informs the systematic analysis of reported UFO incidents. This process aims to construct testable hypotheses and seek demonstrable evidence to substantiate extraordinary claims.

Peer Review and Cross-Disciplinary Discourse: The involvement of peer review within the scientific community, as well as cross-disciplinary dialogue involving experts in astronomy, atmospheric science, psychology, engineering, and other relevant fields, supports the critical assessment of alleged UFO encounters.

Global Collaboration and Experimentation:
International coordination, collaboration, and experimentation among scientists, observatories, and

research institutions foster a collective effort in the scientific scrutiny of alleged UFO incidents. This collaboration contributes to a collective approach in examining and testing reported claims.

Transparency and Publication of Findings: The transparent dissemination of investigative processes, methodological approaches, and findings from scientific scrutiny enhances the credibility and accountability of UFO investigations. Scientific scrutiny of alleged encounters involves transparent communication and publication of results within the scholarly community.

By applying these principles, the scientific community aims to maintain a balanced, evidence-based perspective on UFO encounters, avoiding unwarranted speculation while addressing public interest and legitimate empirical investigation of reported phenomena.

Government transparency and the release of classified

The topic of government transparency and the release of classified information is a complex and multifaceted issue that intersects with considerations of national security, public interest, accountability, and historical disclosure. The release of classified information involves a rigorous and deliberative process, guided by legal frameworks, security protocols, and the principles of transparent governance. Key facets of this topic include:

National Security Imperatives: Government decisions related to the release of classified information are informed by national security imperatives, including the protection of sensitive intelligence, defense capabilities, diplomatic relations, and counterintelligence protocols. Balancing transparency with national security considerations is a fundamental aspect of this process.

Declassification Procedures: The declassification of information involves established procedures and review mechanisms, where designated authorities evaluate the potential risks, historical significance, and public interest considerations involved in releasing classified documents. This process aims to allow for the responsible disclosure of information while safeguarding sensitive data.

Public Interest and Access to Information: Transparency and the release of certain classified information support public interest and the public's right to access government records that are deemed appropriate for declassification. The release of historical documents, records of previous government activities, and information on matters of significance to the public plays a role in promoting transparency and accountability.

Historical Disclosure and Accountability: The release of classified information contributes to historical disclosure, fostering accountability, understanding of past events, and the documentation of governmental decisions, policies, and actions. This can inform public discourse, academic research, and societal understanding of historical events.

Protection of Sensitive Sources and Methods: Classified information often pertains to intelligence sources, collection methods, ongoing operations, and sensitive diplomatic negotiations. The responsible release of information involves safeguarding these sources and methods while balancing the public's interest in transparency.

Ethical Considerations and Privacy Concerns: Ethical considerations, including the protection of individual privacy, intelligence assets, and the diligence of disclosure to prevent harm to individuals or national security, inform decisions about the release of classified information.

Legal Frameworks and Oversight: The release of classified information is subject to legal frameworks, oversight mechanisms, and checks and balances to ensure responsible declassification that complies with relevant laws, executive orders, and regulations.

Selective Redaction and Contextual Disclosure: In cases where complete disclosure is not possible, selective redaction, contextual disclosure, and restricted access to sensitive details may be applied to balance transparency with national security considerations.

Overall, government transparency and the release of classified information involve a delicate balance between public interest, national security imperatives, historical disclosure, and ethical considerations. The complexities inherent in this issue necessitate a careful and comprehensive approach to responsible transparency and the safeguarding of sensitive information.

Critical thinking when approaching UFO evidence

Critical thinking when approaching evidence related to unidentified flying objects (UFOs) is essential for evaluating claims, scrutinizing reported incidents, and discerning the nature of observed phenomena within a framework of skepticism, empirical analysis, and reasoning. The application of critical thinking principles in the examination of UFO evidence involves several key aspects:

Analyzing Observational Data: Examination of observational data, including eyewitness accounts, photographs, videos, and sensor readings, involves a critical assessment of the reliability, credibility, and consistency of reported incidents. This analysis aims to discern the quality of evidence and identify key discrepancies or points of corroboration.

Assessing Natural Explanations: Critical thinking involves considering natural atmospheric phenomena, astronomical objects, human-made aircraft, drones, or other terrestrial sources as potential explanations for observed UFO incidents. This process aims to rule out conventional causes before exploring extraordinary claims.

Questioning Assumptions and Biases: Critical thinking challenges assumptions, cognitive biases, and preconceptions that may influence the interpretation of UFO evidence. This process entails an open-minded approach that refrains from undue speculation or unwarranted belief in extraordinary claims.

Validating Physical Evidence: Rigorous examination of physical evidence, including materials, residues, or traces allegedly associated with UFO incidents, involves a critical evaluation of the authenticity, provenance, and scientific basis of claimed artifacts.

Considering Psychosocial Factors: Critical thinking includes the consideration of psychosocial factors, human cognition, memory, perception, and psychological influences that may affect eyewitness testimony and memory recall in reported UFO encounters. This analysis informs the assessment of the reliability of observational accounts.

Demanding Empirical Verification: The application of critical thinking principles involves demanding empirical verification, testable evidence, and scientifically sound methodology to substantiate extraordinary UFO claims. This approach aims to avoid unwarranted belief in anecdotal or unverified accounts.

Weighing Extraordinary Claims: Critical thinking requires the scrutiny of extraordinary claims, including alleged extraterrestrial visitation or advanced technological phenomena, with a balanced and evidence-based approach that prioritizes empirical verification and falsifiability.

Encouraging Skepticism and Inquiry: Embracing skepticism, open inquiry, and reasoned skepticism fosters a critical yet objective approach to evaluating

UFO evidence, avoiding undue credulity while refraining from categorical dismissal of reported phenomena.

By applying critical thinking principles, individuals, researchers, and the public can approach UFO evidence in a manner that promotes reasoned inquiry, empirical rigor, and an evidence-based understanding of reported incidents. This approach serves to inform public discourse, scientific investigation, and societal understanding of unexplained aerial phenomena.

Through the Sci-Fi Lens: Cultural Impact of Alien Existence

Through the sci-fi lens, the cultural impact of alien existence is a multifaceted and thought-provoking exploration of humanity's fascination with the possibility of extraterrestrial life. Sci-fi literature, films, and media have played a significant role in shaping public imagination, reflecting societal attitudes, and contributing to the enduring cultural impact of the concept of alien existence. Here are some key aspects of the cultural impact of alien existence through the sci-fi lens:

Inspiration and Imagination: Sci-fi narratives featuring alien encounters inspire imagination, creativity, and the exploration of the unknown, fostering a sense of wonder and speculation about the potential diversity of life beyond Earth.

Exploration of Existential Questions: The concept of alien existence in sci-fi provides a platform for exploring existential questions, including humanity's place in the universe, the nature of consciousness, and the potential for interconnectedness with other intelligent beings.

Social Commentary and Allegory: Sci-fi stories often serve as a vehicle for social commentary and allegory, addressing contemporary societal, political, and cultural issues through metaphorical representations of alien encounters, interstellar conflicts, and diplomatic exchanges.

Technological Innovation and Speculation: Sci-fi depictions of alien civilizations stimulate speculation and innovation in science, technology, and space exploration, inspiring the development of advanced technologies and the pursuit of interstellar studies.

Cultural Diversity and Interspecies Relations: Sci-fi narratives featuring alien characters or cultures promote reflection on diversity, tolerance, and the potential for interspecies relations, inviting discussions on cross-cultural understanding and cooperation.

Ethical and Philosophical Contemplation: The concept of extraterrestrial life in sci-fi prompts ethical and philosophical contemplation of the implications of alien existence, the nature of intelligence, and the impact of potential contact with other civilizations.

Fear and Fascination: Sci-fi representations of alien beings evoke a range of emotions, including fear, fascination, awe, and curiosity, reflecting humanity's ambivalent attitudes toward the unknown and the potential for extraterrestrial encounters.

Cultural Icons and Archetypes: Iconic depictions of aliens in popular culture, including portrayals of alien species, creatures, and intelligent beings, have become enduring archetypes that shape public discourse and influence perceptions of potential extraterrestrial life.

By engaging with the concept of alien existence through the sci-fi lens, individuals and societies explore the boundaries of imagination, probe existential questions, and grapple with the potential impact of encountering other intelligent beings, contributing to a rich tapestry of cultural narratives, contemplation, and speculation about the potential diversity of life in the cosmos.

Science fiction's role in shaping public opinion

Science fiction plays a significant role in shaping public opinion by inspiring contemplation, reflection, and conversation on a broad array of scientific, ethical, and societal themes. As a reflection of human imagination, creativity, and speculative inquiry, science fiction narratives influence public perception, attitudes, and cultural perspectives in several key ways:

Imaginative Exploration: Science fiction engages audiences with imaginative explorations of future technologies, extraterrestrial encounters, societal advancements, and speculative concepts, fostering a sense of wonder and curiosity about the potential frontiers of scientific discovery.

Social Commentary and Reflection: Sci-fi narratives often serve as allegorical reflections of contemporary societal, political, and ethical issues, stimulating public dialogue, introspection, and critical examination of cultural norms and values.

Influence on Technological Aspirations: Science fiction narratives inspire technological aspirations, scientific innovation, and the pursuit of advancements in space exploration, robotics, artificial intelligence, and other fields, shaping public enthusiasm and support for scientific endeavors.

Diverse Representation and Inclusivity: Sci-fi narratives contribute to diverse representation, inclusivity, and the exploration of alternative perspectives, fostering public awareness, empathy, and cultural understanding of diverse individuals, societies, and potential extraterrestrial civilizations.

Ethical Contemplation and Philosophical Inquiry: Science fiction prompts ethical contemplation, philosophical inquiry, and reflection on the human condition, inviting public consideration of existential questions, moral dilemmas, and the implications of scientific and technological progress.

Discovery of New Ideas and Frontiers: Through the portrayal of imaginative scenarios, alien encounters, alternative futures, and speculative concepts, science fiction introduces audiences to new ideas, frontiers of exploration, and imaginative possibilities, inspiring public interest in scientific, cosmic, and cultural realms.

Fostering an Informed Citizenry: By igniting curiosity, critical thinking, and informed discourse, science fiction contributes to a more engaged and informed citizenry, catalyzing public interest in science, space exploration, and technological advancements.

Reflection on Existential Questions: Science fiction narratives prompt public reflection on existential questions, the search for meaning, the human experience, and the potential for interstellar contact, stimulating contemplation of the place of humanity in the cosmos and the nature of intelligent life.

By providing a platform for speculative inquiry, imaginative storytelling, and contemplation of the unknown, science fiction narratives shape public opinion, inspire scientific curiosity, and foster cultural dialogue, contributing to a broader societal understanding of the frontiers of knowledge, ethics, and the potential impact of advancements in science and technology.

Impact of movies and literature on our perception of aliens

Movies and literature have a profound and multifaceted impact on our perception of aliens, shaping societal attitudes, speculative inquiry, and cultural fascination with the concept of extraterrestrial life. Through imaginative storytelling, visual representation, and narrative exploration, these mediums influence public perception of aliens in several key ways:

Cultural Archetypes and Iconography: Movies and literature create iconic representations of alien beings, fostering enduring archetypes, symbolism, and visual concepts that shape public perception of extraterrestrial life.

Humanoid and Non-Humanoid Depictions: The portrayal of humanoid and non-humanoid aliens in movies and literature influences public imagination and speculation about the potential diversity of extraterrestrial civilizations, fostering contemplation of alternative forms of life.

Fear and Fascination: The depiction of aliens in various genres evokes a range of emotions, including fear, awe, curiosity, and fascination, influencing societal attitudes and emotional responses toward the concept of encountering intelligent beings.

Technological Advancements and Behaviors: Science fiction narratives explore advanced technologies, cultural behaviors, communication methods, and societal structures of alien civilizations, prompting reflection on the potential diversity of technological and cultural development in the cosmos.

Interspecies Communication and Relations: Movies and literature portray various scenarios of interspecies communication, diplomatic exchanges, and interstellar relations, stimulating public discourse about the potential for intercultural understanding and cooperation with extraterrestrial beings.

Narrative Exploration of Ethical and Moral Themes: Sci-fi narratives offer ethical and moral exploration of human-alien interactions, addressing questions of empathy, tolerance, and the ethical treatment of alien beings, influencing public contemplation of ethical implications.

Social Commentary and Cultural Reflection: Alien narratives serve as allegorical reflections of contemporary societal, political, and ethical issues, providing narratives that prompt critical reflection, social commentary, and cultural introspection on diverse human and alien societies.

Scientific Curiosity and Speculative Inquiry: Movies and literature inspire scientific curiosity, speculative inquiry, and popular interest in cosmic exploration, prompting public enthusiasm for space science, astrobiology, and the search for extraterrestrial life.

By engaging with imaginative narratives and visual representations of aliens, movies and literature influence public perception, inspire scientific inquiry, and foster cultural dialogue on the potential diversity, behaviors, and societal implications of encountering extraterrestrial life, enriching societal understanding of the frontiers of scientific discovery, human imagination, and the quest to unravel the cosmic enigma of alien existence.

How fictional narratives can influence scientific research

Fictional narratives influence scientific research by inspiring curiosity, shaping scientific discourse, and stimulating speculation about the frontiers of scientific knowledge. These narratives foster creativity, conceptual exploration, and interdisciplinary dialogue, leading to impact in several key ways:

Inspiration for Inquiry: Fictional narratives inspire scientific curiosity, prompting researchers to explore speculative concepts, alternative possibilities, and unconventional avenues of inquiry in the pursuit of scientific discovery.

Exploration of Hypothetical Scenarios: Sci-fi narratives create hypothetical scenarios, speculative dilemmas, and imaginative thought experiments that encourage reflection on scientific questions, ethical considerations, and societal implications of scientific advancements.

Conceptual Innovation and Ideation: Fictional narratives stimulate conceptual innovation, ideation, and the exploration of unconventional ideas that influence the development of scientific hypotheses, research methodologies, and experimental approaches.

Promotion of Interdisciplinary Collaboration: Sci-fi narratives foster interdisciplinary collaboration, drawing together expertise from diverse fields—such as science, technology, engineering, arts, and mathematics (STEAM)—to explore imaginative concepts and speculative scenarios.

Stimulation of Technological Advancements: Fictional narratives inspire technological advancements and concept development, stimulating interest in innovative technologies, engineering solutions, and scientific exploration rooted in imaginative possibilities.

Promotion of Public Engagement: Engaging fictional narratives drive public interest, outreach, and participation in scientific dialogue, fostering enthusiasm for scientific pursuits, space exploration, and related areas of research.

Influence on Ethics and Moral Considerations: Sci-fi narratives prompt reflection on ethical and moral dimensions of scientific advancements, influencing contemplation of the societal implications, possible consequences, and responsible stewardship of scientific research.

Shaping Scientific Imagination and Ingenuity: Fictional narratives influence scientific imagination, ingenuity, and the inclination to explore unconventional ideas, encouraging researchers to contemplate the boundaries of scientific understanding and the frontiers of potential discovery.

By engaging with speculative narratives, scientists are inspired to explore the unknown, challenge preconceptions, and envision alternative perspectives that drive scientific innovation, interdisciplinary collaboration, and the expansion of human understanding of the cosmos. This dialogue between fiction and scientific research fosters a dynamic relationship that enriches scientific discourse, stimulates curiosity, and influences the pursuit of knowledge.

The balance between science fiction and science fact

The balance between science fiction and science fact encompasses a dynamic relationship that intertwines imaginative speculation with empirical inquiry, prompting exploration, inspiration, and interdisciplinary dialogue at the intersection of fiction and scientific research. This balance involves several key considerations:

Imaginative Speculation and Empirical Rigor: Science fiction engages in imaginative speculation about the frontiers of scientific knowledge, prompting curiosity, creativity, and conceptual exploration. This imaginative speculation complements empirical rigor, serving as a source of inspiration for scientific research and technological innovation.

Inspiration for Scientific Inquiry: Science fiction narratives inspire scientific curiosity, prompting researchers to explore speculative concepts, alternative possibilities, and unconventional avenues of empirical investigation in the pursuit of scientific discovery.

Exploration of Ethical and Societal Implications: Science fiction narratives promote reflection on ethical and societal implications of scientific advancements, fostering critical contemplation, interdisciplinary dialogue, and public engagement with the nexus of science and culture.

Innovation and Technological Advancements: Sci-fi narratives inspire technological advancements and concept development, stimulating interest in innovative technologies, engineering solutions, and scientific exploration rooted in imaginative possibilities, driving the balance between speculative possibilities and applied research.

Scientific Creativity and Conceptual Innovation: The speculative nature of science fiction encourages scientific creativity, ideation, and the exploration of unconventional ideas, influencing the development of scientific hypotheses, research methodologies, and experimental approaches.

Influence on Public Engagement and Outreach: Science fiction narratives drive public interest, outreach, and engagement with science, fostering enthusiasm for scientific pursuits, space exploration, and the search for extraterrestrial life, shaping public understanding and enthusiasm for scientific research.

Speculative Inquiry and Empirical Testing: Science fiction narratives prompt speculative inquiry, offering thought experiments and hypothetical scenarios, which complement the empirical testing, verification, and validation processes inherent in scientific research.

Human Imagination and the Quest for Knowledge: The balance between science fiction and science fact reflects the dynamic interplay between human imagination and the quest for empirical knowledge, encompassing a diverse and rich dialogue that enriches scientific creativity, ingenuity, and interdisciplinary exploration.

By navigating the intersection of science fiction and science fact, researchers, writers, and creative minds engage in a dynamic dialogue that fosters an equilibrium between imaginative speculation and empirical inquiry, contributing to scientific advancements, cultural reflection, and the exploration of the unknown. This balance represents a symbiotic relationship that enriches both the creative realms of fiction and the pursuit of knowledge

Ethical and Philosophical Implications of Contact

The ethical and philosophical implications of potential contact with extraterrestrial civilizations prompt contemplation of far-reaching questions, moral considerations, and existential reflections on the nature of human interaction with other intelligent beings. These implications intersect with ethical frameworks, societal values, and philosophical musings, shaping discourse, speculation, and introspection about the potential impact of interstellar contact. Here are key aspects of the ethical and philosophical implications of contact with extraterrestrial civilizations:

Cosmic Stewardship and Responsibility: Consideration of the ethical dimensions of contact involves contemplation of humanity's role as custodians of the cosmos, fostering dialogue about stewardship, responsibility, and the ethical use of scientific advancements in interstellar communication and exploration.

Encounter of Otherness and Diversity: Potential contact with extraterrestrial civilizations raises questions about the encounter of otherness, cultural diversity, and interspecies relations, stimulating reflection on the ethical treatment, respect, and understanding of alien beings.

Implications for Humanity's Identity and Worldview: The prospect of interstellar contact prompts contemplation of humanity's identity, worldview, and self-conception in the context of the broader cosmos, influencing philosophical inquiry into human purpose, existence, and the place of intelligent life in the universe.

Technological and Moral Dilemmas: Ethical considerations encompass the potential technological and moral dilemmas associated with interstellar communication, including discussions about the impact of advanced knowledge, the implications of technological sharing, and the potential for cooperation or conflict.

Cultural, Societal, and Religious Perspectives: Reflection on the ethical and philosophical implications of contact encompasses considerations of diverse cultural, societal, and religious perspectives, fostering dialogue about the moral dimensions, existential questions, and societal implications of potential interstellar relations.

Ethical Practice and Intercultural Understanding: Ethical considerations prompt reflection on the value of ethical practice, intercultural understanding, and the potential for cooperative, peaceful, and respectful engagement with extraterrestrial civilizations.

Implications for Scientific Advancement and Human Progress: Philosophical implications of contact intersect with the advancement of scientific knowledge, the pursuit of human progress, and the potential impact on society, influencing discourse on the ethical dimensions of scientific inquiry and societal advancement in the context of cosmic exploration.

Long-Term Societal and Planetary Implications: The long-term ethical and philosophical implications of potential contact prompt consideration of the societal, planetary, and existential implications of interstellar relations, influencing discourse on the impact of extraterrestrial contact on human affairs, societal structures, and planetary stewardship.

By contemplating the ethical and philosophical implications of potential contact with extraterrestrial civilizations, individuals, societies, and the scientific community engage in substantive dialogue, introspection, and speculative inquiry that enriches ethical frameworks, inspires philosophical reflection, and promotes cultural awareness of the broader implications of interstellar engagement.

The potential impact of discovering extraterrestrial intelligence

The potential impact of discovering extraterrestrial intelligence encompasses a wide array of implications, ranging from scientific, sociocultural, and philosophical consequences to ethical, technological, and societal considerations. The discovery of extraterrestrial intelligence holds the potential to profoundly influence humanity's understanding of itself, the cosmos, and the implications of encountering other intelligent beings. Here are key aspects of the potential impact of discovering extraterrestrial intelligence:

Scientific and Philosophical Paradigm Shift: The discovery of extraterrestrial intelligence has the potential to trigger a paradigm shift in scientific and philosophical inquiry, reshaping humanity's understanding of the universe, existence, and the nature of intelligent life.

Technological Advancements and Innovation: The search for and discovery of extraterrestrial intelligence inspire scientific and technological advancements, stimulating innovation in space exploration, astrobiology, communication technologies, and the pursuit of advanced knowledge.

Ethical and Moral Contemplation: Discovery engenders ethical and moral contemplation about humanity's role as cosmic stewards, prompting reflection on the ethical dimensions, moral responsibilities, and potential impact of interstellar contact.

Sociocultural Reflection and Intercultural Understanding: The discovery of extraterrestrial intelligence fosters sociocultural reflection, promoting an appreciation of interspecies diversity, encouraging intercultural understanding, and shaping public discourse about the nature of intelligent life in the cosmos.

Cultural, Artistic, and Literary Expression: The impact of discovery resonates in cultural, artistic, and literary expression, fueling creative imagination, inspiring storytelling, and contributing to societal contemplation of the implications of encountering other intelligent beings.

Existential Questions and Human Identity: Discovery prompts contemplation of existential questions, influencing societal understanding of human identity, purpose, and existence in the context of the broader universe and the potential for interstellar contact.

Public Engagement and Awareness: The impact of discovery stimulates public engagement, awareness, and enthusiasm for scientific inquiry, space exploration, and technological advancements, shaping public interest in cosmic exploration and the search for other intelligent beings.

Scientific and Societal Collaboration: The potential impact fosters collaboration among scientists, societal leaders, policymakers, and the public, encouraging dialogue, cooperation, and international engagement in understanding the implications of extraterrestrial intelligence.

By considering the potential impact of discovering extraterrestrial intelligence, individuals, societies, and the scientific community engage in reflection, dialogue, and speculative inquiry that enriches cultural awareness, scientific discourse, and the pursuit of knowledge about the broader implications of contact with other intelligent beings.

Ethical considerations in alien contact scenarios

In the hypothetical scenario of contact with extraterrestrial intelligence, ethical considerations encompass a wide array of moral, societal, and intercultural implications that prompt contemplation of responsible engagement, stewardship, and the ethical dimensions of interspecies interaction. These considerations influence discourse, ethical frameworks, and the potential for societal, philosophical, and existential implications in the event of interstellar contact. Here are key ethical considerations in alien contact scenarios:

Ethical Frameworks and Interspecies Relations: Ethical considerations prompt reflection on the development of ethical frameworks, moral principles, and the potential for intercultural understanding, fostering dialogue about the responsible treatment and respect for other intelligent beings.

Cultural Sensitivity and Intercultural Exchange: Ethical contemplation encompasses an appreciation of cultural diversity, societal values, and the potential for intercultural exchange and cooperation, encouraging respect, sensitivity, and understanding in human-alien interactions.

Responsible Communication and Interaction: The ethical considerations of contact prompt reflection on responsible communication, the avoidance of harm, and the promotion of peaceful interactions, guiding the ethical practice of scientific, diplomatic, and societal engagement with extraterrestrial civilizations.

Planetary Stewardship and Interstellar Diplomacy: Ethical implications prompt contemplation of planetary stewardship, interstellar diplomacy, and the potential impact of intercultural relations on societal structures, environmental preservation, and the shared responsibilities of cosmic stewardship.

Inclusive Engagement and Global Collaboration: Ethical dimensions foster inclusive engagement, global collaboration, and international dialogue about societal, ethical, and cultural considerations of interspecies engagement, encouraging the development of ethical guidelines and intergovernmental cooperation.

Societal Awareness and Preparedness: Ethical considerations prompt reflection on societal awareness, preparedness, and the dissemination of information about the potential implications of contact with extraterrestrial intelligence, promoting a responsible approach to public engagement and outreach.

Ethical Practice and Moral Dilemmas: Ethical implications prompt contemplation of moral dilemmas, ethical decision-making, and the responsible practice of scientific inquiry, fostering reflection on the potential consequences and societal implications of interstellar contact.

Environmental and Societal Implications: Ethical considerations encompass the potential environmental, societal, and existential implications of interspecies interaction, influencing discourse about the impact on human affairs, societal structures, and planetary stewardship.

By considering these ethical implications, individuals, societies, and the scientific community engage in substantive dialogue, introspection, and speculative inquiry that enriches ethical practice, inspires philosophical reflection, and fosters a thoughtful, responsible approach to the potential implications of extraterrestrial contact

Philosophical questions raised by the existence of alien life

The existence of alien life raises a multitude of thought-provoking philosophical questions that touch on the fundamental nature of humanity, the cosmos, and the nature of intelligence. These queries prompt contemplation of existential, ethical, and epistemological themes, stimulating reflection on the broader implications of extraterrestrial life. Here are some key philosophical questions raised by the existence of alien life:

The Nature of Intelligence: The existence of alien life prompts reflection on the nature of intelligence, consciousness, and the potential diversity of cognitive capacities in the cosmos, fostering philosophical inquiry into the boundaries of cognition and the potential for alternative forms of mental experience.

Anthropocentrism and Human Exceptionalism: The existence of alien life challenges anthropocentric perspectives, prompting reflection on human exceptionalism, the uniqueness of human existence, and the potential variation, diversity, and societal organization of intelligent beings in the universe.

Existential Questions and Human Identity: The existence of alien life prompts contemplation of existential questions about human identity, purpose, and existence in the context of the broader universe, influencing philosophical reflection on the nature of human consciousness and the place of humanity in the cosmos.

Ethical and Interspecies Considerations: The existence of alien life raises ethical and intercultural questions about human-alien relations, prompting contemplation of ethical frameworks, the responsible treatment of other beings, and the potential for interspecies cooperation, fostering ethical inquiry into respectful engagement with extraterrestrial life.

Epistemological Reflection: The existence of alien life inspires epistemological reflection on the conditions for knowledge, the limitations of human understanding, and the potential for alternative methodologies, prompting inquiry into the scope and boundaries of human knowledge in the face of encounters with other intelligent beings.

Societal, Political, and Diplomatic Implications: The existence of alien life fosters reflection on societal, political, and diplomatic themes, influencing discourse about the implications for human affairs, societal structures, interstellar diplomacy, and planetary stewardship in the context of potential contact.

Ethical and Cosmic Responsibility: The existence of alien life prompts reflection on the implications for cosmic responsibility, the ethical dimensions of planetary stewardship, and the potential ethical obligations associated with interstellar communication and engagement.

Philosophical Speculation and Imaginative Inquiry: The existence of alien life stimulates philosophical speculation and imaginative inquiry about the nature of intelligent beings, the potential for cosmic awareness, and the societal, ethical, and philosophical implications of extraterrestrial existence.

By engaging with these philosophical questions, individuals, societies, and the scientific community enter into a substantive dialogue that enriches ethical practice, inspires philosophical reflection, and fosters a thoughtful, considered approach to the potential implications of the existence of alien life.

Preparing humanity for the possibility of alien life

Preparing humanity for the possibility of alien life involves fostering a thoughtful, informed approach to the prospect of encountering other intelligent beings, reflecting on our societal, ethical, and philosophical readiness and the potential impact on humanity's understanding of itself and the cosmos. Here are key aspects of preparing humanity for the possibility of alien life:

Scientific Inquiry and Interdisciplinary Dialogue: Encouraging scientific exploration, interdisciplinary collaboration, and open dialogue about the potential implications of extraterrestrial encounters, fostering an informed, evidence-based approach to understanding the cosmos.

Ethical Consideration and Societal Preparedness: Promoting ethical reflection, societal awareness, and the dissemination of information about the potential implications of contact with extraterrestrial intelligence, fostering a responsible, informed approach to public engagement and outreach.

Cultural Sensitivity and Humanistic Values: Encouraging cultural sensitivity, intercultural understanding, and the exploration of humanistic values in the context of potential interstellar engagement, promoting reflection on our societal attitudes, responses, and societal implications of these interactions.

Philosophical Contemplation and Intellectual Inquiry: Stimulating philosophical reflection, intellectual inquiry, and speculative contemplation about the societal, existential, and philosophical implications of potential interstellar contact, fostering a thoughtful, mindful approach to the prospect of encountering other intelligent beings.

Environmental and Planetary Stewardship: Prompting reflection on environmental stewardship, planetary preservation, and the potential societal, environmental, and planetary implications of interstellar interactions, fostering responsible cosmic stewardship and the consideration of the impact on planetary ecosystems.

Public Engagement and Informed Dialogue: Encouraging informed public engagement, public dialogue, and thoughtful discourse about the societal, ethical, and philosophical implications of potential contact with extraterrestrial intelligence to foster a well-informed, enquiring public ready for the prospect of other intelligent beings.

International Collaboration and Diplomatic Preparedness: Fostering international collaboration, diplomatic preparedness, and cross-cultural understanding, promoting global readiness for potential societal, diplomatic, and intercultural implications of encounters with other intelligent beings.

Scientific and Technological Readiness: Encouraging technological preparedness, scientific cooperation, and advancements in communication technologies, inspiring the pursuit of advanced knowledge, technical excellence, and the development of interstellar communication methods in anticipation of potential contact.

By promoting discourse, reflection, and informed engagement on the possibility of alien life, humanity can cultivate a thoughtful, ethical, and informed approach to the prospect of encountering other intelligent beings, fostering preparedness and a mindful approach to the potential implications of extraterrestrial contact.

Emerging Technologies and the Future of Space Exploration

Emerging technologies are poised to revolutionize the future of space exploration, enabling new frontiers of scientific discovery, technological innovation, and human expansion beyond Earth. These technologies span a diverse array of fields, from propulsion and materials science to robotics and artificial intelligence, and are expected to shape the trajectory of space exploration in several key ways:

Advanced Propulsion Systems: Emerging propulsion technologies, such as ion propulsion, solar sails, and plasma thrusters, offer enhanced efficiency and capabilities for deep space exploration, enabling sustained missions to outer planets, asteroids, and interstellar destinations.

In-Situ Resource Utilization: Technologies for in-situ resource utilization, including 3D printing, resource extraction, and manufacturing capabilities, have the potential to enable sustainable human presence on the Moon, Mars, and beyond by leveraging local materials for construction, infrastructure development, and life support systems.

Autonomous Rovers and Robotics: Advancements in autonomous rovers, robotics, and AI-driven exploration systems offer enhanced capabilities for planetary surface exploration, sample collection, and geological analysis, paving the way for enhanced understanding of celestial bodies and the potential for human settlement.

Telepresence and Virtual Reality: Technologies for telepresence and virtual reality are expected to enable immersive, real-time exploration experiences for astronauts and researchers, enhancing our capacity to remotely explore and interact with distant planetary environments and celestial bodies.

Space-Based Telescopes and Observatories: Advanced space-based telescopes and observatories, including the James Webb Space Telescope and future next-generation successors, will offer unprecedented capabilities for astronomical observation, enabling the study of exoplanets, galactic systems, and the early universe.

Advanced Life Support and Bioengineering: Emerging technologies in life support systems, bioengineering, and closed-loop ecological habitats will bolster the resilience and sustainability of crewed missions, facilitating extended stays on planetary surfaces and deep space environments.

Small Satellite and CubeSat Innovations: Advancements in small satellite and CubeSat technologies are expected to democratize access to space, foster cost-effective exploration missions, and enable a more distributed, collaborative approach to space exploration and scientific research.

Interplanetary Communication and Data Relay: Emerging communication and data relay technologies will enable enhanced interplanetary connectivity, facilitating real-time communication with probes, rovers, and human missions across vast distances within the solar system.

By harnessing the potential of these emerging technologies, the future of space exploration is set to be characterized by enhanced scientific discovery, expanded capabilities for human spaceflight, and the pursuit of innovative methodologies for studying, understanding, and exploring the cosmos.

Advancements in space travel and exploration

Advancements in space travel and exploration encompass a diverse array of technologies, methodologies, and mission architectures that are propelling the frontiers of scientific discovery, human spaceflight, and robotic exploration. These advancements are shaping the future of space exploration in several key areas:

Human Spaceflight Capabilities: Advancements in human spaceflight include the development of next-generation spacecraft, life support systems, and habitat modules, fostering the potential for extended crewed missions, lunar exploration, and future human missions to Mars.

Sustainable Lunar Exploration: Initiatives focused on lunar exploration aim to enable sustainable, long-term presence on the Moon through the development of habitats, in-situ resource utilization technologies, and collaborative partnerships for international lunar missions.

Mars Exploration Architecture: Advancements in Mars exploration architecture encompass the development of advanced propulsion systems, entry, descent, and landing technologies, and in-situ resource utilization capabilities to enable future human missions to Mars.

Space Launch Systems: The development of next-generation heavy-lift launch systems, reusable rockets, and modernized launch infrastructures is bolstering the capacity for payload delivery, interplanetary missions, and crewed spaceflight endeavors.

Planetary Rovers and Probes: Advancements in robotic exploration include the development of advanced planetary rovers, landers, and orbital probes equipped with enhanced scientific instrumentation and autonomous capabilities for studying celestial bodies and planetary surfaces.

Deep Space Telescopes and Observatories: Advancements in space telescopes, such as the James Webb Space Telescope and future large-scale observatories, signal the potential for breakthroughs in astronomical observation, enabling unprecedented insights into exoplanets, galactic structures, and the early universe.

Collaborative International Missions: Collaborative international missions and partnerships in space exploration, including joint crewed missions, robotic landers, and deep space probes, offer the potential for shared resources, expertise, and global cooperation in the pursuit of scientific discovery.

Private Sector Innovations: Innovations in the private sector, including commercial spaceflight ventures, space tourism initiatives, and advancements in space infrastructure, are expanding the scope of space exploration capabilities and fostering diverse, multipurpose applications for space travel and research.

By harnessing these advancements in space travel and exploration, the future of space exploration is characterized by enhanced scientific discovery, expanded capabilities for human spaceflight, and the pursuit of innovative methodologies for studying, understanding, and exploring the cosmos.

The rise of private space companies and their contributions

The rise of private space companies has transformed the space industry, leading to notable advancements in space technology, launch capabilities, and commercial applications, dramatically expanding the scope and diversity of space exploration. These companies have made significant contributions in several key areas:

Launch Capabilities: Private space companies, such as SpaceX, Blue Origin, and Rocket Lab, have developed advanced launch vehicles, reusable rocket technology, and modernized launch infrastructure, significantly lowering the cost of access to space and facilitating increased launch frequency.

Space Tourism: Private space companies are pioneering the development of space tourism initiatives, offering the potential for commercial suborbital and orbital spaceflights for private individuals, thus democratizing access to space and inspiring public engagement in space exploration.

Orbital and Suborbital Research Missions: These companies are enabling commercial access to microgravity research, technology demonstrations, and commercial payloads through orbital and suborbital missions, opening new avenues for scientific research and industrial applications in space.

Satellite Constellations and Telecommunications: Private space companies are leading the development of advanced satellite constellations for global broadband internet access, Earth observation, and telecommunication services, fostering innovative applications and global connectivity.

Lunar and Planetary Exploration: Private companies are actively engaged in the development of lunar and planetary exploration missions, including landers, rovers, and interplanetary spacecraft, contributing to the advancement of robotic exploration and potential resource utilization on the Moon and beyond.

International and Public-Private Partnerships: Private space companies are fostering collaborations with international space agencies, public-private partnerships, and shared commercial ventures, thereby expanding the scope of space exploration and furthering cooperative efforts in scientific research, technological innovation, and societal applications.

Space Infrastructure and In-Space Services: Advanced concepts in space infrastructure, in-space manufacturing, servicing, and resource utilization are being pursued by private space companies, with the potential to enhance the sustainability and scalability of human activities in space.

Technological Innovation and Cost Reduction: Private companies are driving technological innovation and cost reduction in space missions through the application of modernized engineering, manufacturing practices, and agile development methodologies, leading to increased efficiency and reduced mission costs.

The contributions of private space companies are reshaping the landscape of space exploration, stimulating new possibilities for research, commerce, and human expansion in space while demonstrating the potential for innovative models of collaboration and entrepreneurship in the space industry.

The role of artificial intelligence in searching for life

Artificial intelligence (AI) plays a crucial role in the search for life beyond Earth, contributing to the analysis of vast datasets, the interpretation of complex signals, and the identification of potential biosignatures. The application of AI in astrobiology and the exploration for extraterrestrial life encompasses several key areas:

Data Analysis and Pattern Recognition: AI algorithms are employed to analyze large volumes of astronomical data, such as spectroscopic measurements or radio signals, enabling pattern recognition and the identification of anomalous or potentially meaningful signals that may indicate the presence of extraterrestrial life.

Identification of Biosignatures: AI assists in the identification of biosignatures—indicators of potential life or habitability—in data gathered from exoplanet observations, atmospheric analyses, and planetary surface investigations, aiding in the quest to discern the potential presence of life beyond Earth.

Autonomous Exploration and Robotic Probes: AI-driven autonomous systems control robotic explorers and probes, enabling adaptive decision-making, hazard avoidance, and prioritization of scientific targets, thus fostering more efficient and responsive exploration of environments that may harbor signs of life.

Language Processing and Intelligent Communication: AI facilitates intelligent language processing and the potential for interstellar communication, aiding in the interpretation of potential extraterrestrial messages or the development of protocols for human-initiated communication with hypothetical alien civilizations.

Data Mining and Information Integration: AI technologies are employed to mine vast databases, integrate diverse sources of information, and perform complex data fusion, enabling comprehensive analysis and interpretation of multi-modal data relevant to the search for extraterrestrial life.

Machine Learning for Planetary Analysis: Machine learning techniques are utilized to analyze planetary characteristics, geological formations, and environmental parameters, helping to identify regions with the potential to support life, and guiding the selection of target sites for future exploration.

Next-Generation Telescopes and Observatories: AI-driven control systems and image processing algorithms are deployed to enhance the capabilities of next-generation telescopes and observatories, enabling advanced data capture, real-time analysis, and the detection of subtle phenomena relevant to the search for life.

Complex Signal Interpretation: AI technologies aid in the interpretation of complex signals, including radio emissions, optical transmissions, and spectral data, enabling the discernment of potential artificial signals or signatures indicative of technological civilizations.

By harnessing the potential of AI, researchers and space agencies are augmenting the capacity for comprehensive analysis, autonomous exploration, and the interpretation of data relevant to the search for extraterrestrial life, paving the way for innovative and responsive approaches to the cosmic quest for understanding the potential for life beyond Earth.

Anticipating the next big discovery in space exploration

Anticipating the next big discovery in space exploration encompasses a wide array of potential breakthroughs, scientific advancements, and technological accomplishments that hold the promise of transforming our understanding of the universe and the potential for human expansion beyond Earth. Several areas of space exploration hold promise for significant discoveries:

Detection of Exoplanets with Biosignatures: The discovery of an exoplanet with compelling biosignatures, such as the presence of atmospheric gases indicative of life, has the potential to revolutionize our understanding of habitable worlds and the potential for extraterrestrial life elsewhere in the universe.

Confirmation of Extraterrestrial Microbial Life: The detection of microbial life beyond Earth, whether within our own solar system or on distant celestial bodies, would represent a groundbreaking discovery with profound implications for our understanding of the potential for life in space.

Evidence of Past or Present Habitability on Mars: Discoveries indicating past or present habitability on Mars, including the confirmation of ancient water features, subsurface reservoirs, or organic molecules, stand to offer significant insights into the potential for Martian life and the history of the Red Planet.

Direct Imaging of Exoplanetary Surfaces or Oceans: The direct imaging of exoplanetary surfaces, oceans, or landforms on distant worlds using advanced telescopes and observational techniques would represent a transformative leap in our capacity to study potentially habitable exoplanets.

Technosignatures of Technological Civilizations: Evidence of technosignatures, such as artificial structures, electromagnetic emissions, or extraterrestrial technology, would represent a major discovery indicating the potential presence of technological civilizations beyond Earth.

Exploration Beneath the Icy Moons: Discoveries related to the subsurface oceans and potential habitability of icy moons in our solar system, such as Europa, Enceladus, or Ganymede, could provide key insights into the potential for life beyond Earth.

Unprecedented Insights from Space Telescopes: Next-generation telescopes, such as the James Webb Space Telescope and the upcoming Nancy Grace Roman Space Telescope, have the potential to unveil unprecedented insights into the cosmic dawn, exoplanet atmospheres, and the early universe.

Lunar and Asteroidal Resource Utilization: Breakthroughs in lunar and asteroidal resource utilization, such as the discovery of economically viable resources or volatiles, could have significant implications for the feasibility of long-term human exploration and settlement in space.

By anticipating and pursuing these potential discoveries in space exploration, researchers, space agencies, and the global scientific community are shaping the trajectory of future advancements, expanding the horizons of human knowledge, and nurturing the potential for transformative breakthroughs that will usher in a new era of cosmic understanding.

The Endless Enigma: Embracing the Universe's Mysteries

"**The Endless Enigma**: Embracing the Universe's Mysteries" encapsulates the profound and ongoing dialogue surrounding the enigmatic frontiers of space, the inexhaustible realm of cosmic inquiry, and the enduring pursuit of understanding the cosmos. This evocative theme speaks to the timeless contemplation of the unknown, the infinite possibilities of discovery, and the humbling, ever-expanding journey of human exploration. Here are some elements that can be included in the exploration of this theme:

The Boundless Cosmos: The theme of "The Endless Enigma" captures the grandeur and enigmatic nature of the universe, inviting contemplation of celestial mysteries, cosmic phenomena, and the insatiable curiosity that drives the quest for understanding the cosmos.

Unsolved Cosmic Riddles: Embracing the universe's mysteries encompasses the contemplation of unsolved cosmic riddles, including the nature of dark matter and dark energy, the origins of life, the potential for extraterrestrial intelligence, and the enigmatic properties of black holes.

Philosophical Reflection: The theme fosters philosophical reflection on humanity's place in the cosmos, existential questions, and the quest for meaning, delving into the profound implications of encountering the unknown and the mysteries that defy simple explanation.

Scientific Contemplation: "The Endless Enigma" stimulates scientific inquiry and contemplation of cosmic phenomena, inspiring the pursuit of knowledge, the quest for scientific discovery, and the incomparable wonders of the universe that continue to elude complete understanding.

Imagination and Speculative Inquiry: The theme invites imaginative speculation, speculative inquiry, and the exploration of alternative scenarios, prompting contemplation of the potential frontiers of scientific knowledge and the enigmatic possibilities of the cosmos.

Historical and Cultural Significance: Embracing the universe's mysteries encompasses the historical significance of cosmic inquiry, the cultural impact of celestial wonders, and the enduring human fascination with the unknown, inspiring artistic, literary, and cultural expression.

Public Engagement and Outreach: The theme fosters public engagement, scientific outreach, and education about the enigmatic nature of the universe, inspiring curiosity, wonder, and appreciation for the enduring mysteries of the cosmos.

Innovative Approaches and Advanced Technologies: Embracing the universe's mysteries encapsulates innovative approaches, advanced technologies, and the frontiers of scientific exploration that drive the ongoing pursuit of unraveling enigmatic cosmic phenomena.

By embracing "The Endless Enigma," individuals, researchers, and the public engage in a timeless and thought-provoking dialogue about the mystique of the cosmos, the enigmatic frontiers of scientific inquiry, and the enduring pursuit of understanding the boundless mysteries of the universe.

Understanding the significance of unanswered questions

Understanding the significance of unanswered questions encompasses a profound contemplation of the enigmatic, the enduring, and the thought-provoking mysteries that drive human inquiry and the quest for knowledge. Acknowledging the significance of unanswered questions in science, philosophy, and human understanding involves several key dimensions:

Catalyst for Curiosity and Exploration: Unanswered questions serve as a catalyst for curiosity, sparking exploration, inquiry, and the pursuit of understanding, propelling the frontiers of scientific discovery and human imagination.

Drivers of Scientific Inquiry: Unanswered questions stimulate scientific inquiry, inspiring researchers, thinkers, and visionaries to explore, question, and seek solutions that expand the boundaries of human knowledge and understanding.

Humility and Intellectual Curiosity: Acknowledging unanswered questions fosters humility, intellectual curiosity, and the recognition of the vastness of human understanding, prompting reflection on the unknown and the enduring mysteries of the universe.

Inspiration for Imagination and Creativity: Unanswered questions inspire imagination, creativity, and speculative inquiry, prompting the exploration of alternative scenarios, new ideas, and innovative approaches to unraveling enigmatic phenomena.

Philosophical Reflection and Intellectuability: Unanswered questions foster philosophical reflection, intellectual humility, and contemplation of the enduring mysteries of existence, prompting inquiry into the nature of knowledge, the frontiers of understanding, and the enigmatic possibilities of the cosmos.

Scientific Progress and Technological Innovation: Unanswered questions drive scientific progress and technological innovation, inspiring the development of methodologies, technologies, and advances that contribute to the pursuit of solutions and discoveries in science and engineering.

Cultural, Artistic, and Literary Expression:
Acknowledging unanswered questions inspires cultural, artistic, and literary expression, fostering creative explorations of cosmic mysteries, existential contemplation, and the enduring human fascination with the unknown.

Inclusivity and Collaborative Inquiry: Unanswered questions prompt inclusivity and collaborative inquiry, fostering engagement, dialogue, and diverse perspectives that contribute to the pursuit of understanding and the sharing of knowledge.

By understanding the significance of unanswered questions, individuals, societies, and the scientific community engage in a thoughtful, open-ended dialogue that fosters intellectual curiosity, imaginative exploration, and the enduring pursuit of unraveling cosmic enigmas, advancing the frontiers of human understanding.

The importance of continued exploration and curiosity

Continued exploration and curiosity are of paramount importance to the advancement of human knowledge, the enrichment of cultural understanding, and the nurturing of scientific discovery. The significance of sustained exploration and curiosity encompasses several key dimensions:

Expansion of Human Understanding: Continued exploration and curiosity propel the expansion of human understanding, fostering a dynamic, ongoing dialogue about the natural world, cosmic phenomena, and the quest for knowledge.

Inspiration for Scientific Inquiry: Exploration and curiosity serve as catalysts for scientific inquiry, inspiring researchers, scholars, and thinkers to delve into the enigmatic mysteries of the universe, thus propelling the frontiers of scientific discovery and human imagination.

Cultural Appreciation and Awareness: Continued exploration and curiosity promote cultural appreciation and awareness of the world's diversity, historical significance, and the richness of human experience, fostering intellectual enrichment and societal understanding.

Innovation and Technological Advancements: Curiosity drives innovation and technological advancements, inspiring the development of advanced methodologies, technologies, and solutions that contribute to advancements in science, engineering, and societal progress.

Quest for Solutions and Understanding: The spirit of exploration and curiosity propels the quest for solutions and understanding, driving the pursuit of scientific knowledge, technological innovation, and human endeavor within the natural world and the cosmos.

Philosophical and Intellectual Enrichment: Continued exploration and curiosity enrich philosophical and intellectual contemplation, fostering dialogue, reflection, and speculative inquiry about the enduring mysteries of existence and the frontiers of human knowledge.

Inspiration for Imagination and Creativity: Curiosity inspires imagination, creativity, and speculative inquiry, prompting the exploration of alternative scenarios, new ideas, and innovative approaches to unraveling enigmatic phenomena.

Global Engagement and Collaboration: Exploration and curiosity foster global engagement, dialogue, and collaboration, nurturing inclusive, diverse perspectives and a shared quest for understanding, scientific discovery, and cultural enrichment.

By emphasizing the importance of continued exploration and curiosity, individuals, societies, and the scientific community engage in a dynamic, forward-thinking dialogue that fosters intellectual curiosity, imaginative exploration, and the enduring pursuit of unraveling the mysteries of the universe.

Accepting the vastness and the limits of our understanding

Accepting the vastness and the limits of our understanding engenders humility, curiosity, and a grounded perspective on the frontiers of human knowledge and the mysteries of the cosmos. This acknowledgment encompasses several key dimensions:

Humility in the Face of the Unknown: Accepting the vastness and the limits of our understanding fosters humility, acknowledging the incomprehensible extent of the universe and the enduring mysteries that defy complete elucidation.

Incentive for Curiosity and Exploration: Awareness of the limits of understanding inspires curiosity, exploration, and the quest for knowledge, propelling the frontiers of scientific discovery and intellectual inquiry into the enigmatic mysteries of the universe.

Philosophical Reflection and Humility: Acknowledging the limits of understanding fosters philosophical reflection, intellectual humility, and contemplation of the cosmos, prompting inquiry into the nature of knowledge, the frontiers of understanding, and the enigmatic possibilities of the universe.

Inspiration for Intellectual Inquiry: Awareness of the vastness of the universe and the limits of understanding inspires continued intellectual inquiry, imaginative speculation, and cooperative exploration, nurturing a thoughtful, open-ended approach to the quest for knowledge.

Motivation for Innovative Solutions: Acknowledgement of the limits of understanding motivates the pursuit of innovative solutions, alternative methodologies, and adaptive approaches to unraveling enigmatic phenomena and embracing uncertainty.

Enriching Cultural Reflection: Accepting the vastness and the limits of our understanding enriches cultural reflection, inspiring artistic, literary, and cultural expression, and nurturing a sense of awe, wonder, and contemplation of the enduring mysteries of the cosmos.

Inclusivity and Collaborative Inquiry: Awareness of the limits of understanding and the vastness of the cosmos fosters inclusivity and collaborative inquiry, nurturing engagement, dialogue, and diverse perspectives that contribute to the pursuit of understanding and the sharing of knowledge.

By accepting the vastness and the limits of our understanding, individuals, societies, and the scientific community engage in a thoughtful, open-ended dialogue that fosters intellectual curiosity, imaginative exploration, and the enduring pursuit of unraveling cosmic enigmas, advancing the frontiers of human understanding.

Inspiring future generations to pursue the quest for knowledge

Inspiring future generations to pursue the quest for knowledge encompasses a profound and enduring commitment to nurturing curiosity, intellectual inquiry, and a passion for learning. This shared endeavor is essential for fostering an informed, engaged, and visionary society that propels the frontiers of scientific discovery, cultural enrichment, and human understanding. Here are key dimensions for inspiring future generations:

Cultivating Curiosity and Wonder: Inspiring future generations involves cultivating curiosity, awakening wonder, and fostering a sense of awe for the natural world, the mysteries of the cosmos, and the endless possibilities of human understanding.

Promoting Scientific Literacy and Inquiry: Engaging future generations in scientific literacy, inquiry-based learning, and interactive exploration, fostering a thoughtful, evidence-based approach to understanding the world and the pursuit of reasoned, informed knowledge.

Embracing Intellectual Diversity and Inclusivity: Inspiring future generations involves embracing intellectual diversity, inclusivity, and the value of diverse perspectives that contribute to the quest for knowledge, fostering open dialogue, collaborative inquiry, and shared understanding.

Fostering a Sense of Intellectual Empowerment: Encouraging future generations to pursue the quest for knowledge involves fostering a sense of intellectual empowerment, resilience, and adaptability, nurturing an inquisitive, solution-driven approach to unraveling complex questions and enigmatic phenomena.

Nurturing Passion for Learning and Lifelong Inquiry: Inspiring future generations includes nurturing a passion for learning, critical thinking, and the pursuit of lifelong inquiry, fostering a commitment to intellectual growth, discovery, and the ongoing quest for understanding.

Promoting Cultural Enrichment and Artistic Expression: Inspiring future generations encompasses promoting cultural enrichment, artistic expression, and appreciation for the humanities, fostering imaginative exploration, expressive inquiry, and the cultivation of a holistic understanding of the human experience.

Mentorship and Role Modeling: Encouraging mentorship, support, and positive role modeling from scientific, cultural, and educational leaders, inspiring future generations by sharing personal insights, experience, and a passion for exploration and discovery.

Global Engagement and Interdisciplinary Collaboration: Fostering global engagement and interdisciplinary collaboration, inspiring future generations to embrace diverse perspectives, collaborative inquiry, and shared knowledge that contribute to the ongoing quest for understanding the cosmos.

By inspiring future generations to pursue the quest for knowledge, individuals, educators, and societal leaders cultivate a legacy of intellectual curiosity, imaginative exploration, and enlightened engagement that propels the enduring quest for understanding and the advancement of human knowledge.